Introduction

A: Legal and management

B: Health and welfare

C: General safety

D: High risk activities

E: Environment

Contents

F: Specialist activities

If you are preparing for a specialist test you also need to revise the appropriate specialist activity, from those listed below.

Further information

Foreword

Construction is an exciting industry. There is constant change as work progresses to completion. We all recognise that a healthy and safe workforce is critical to the success of the industry, and it is essential that standards are not compromised.

Sadly, each year the construction industry accounts for more than a quarter of the UK's work-related fatal accidents. Over 4,500 workers die from the long-term effects of breathing in hazardous construction dusts and there are more than 4,000 specified injuries.

Many of these can be avoided by improving workers' hazard and risk awareness, through effective leadership and worker involvement, by providing the right tools and equipment, and a continued commitment to a fully trained and competent workforce.

The CITB *Health, safety and environment test* plays an important role in helping to achieve this. It is a key part of obtaining a competency card. The test continues to be a vital tool in helping the industry to improve its health, safety and environmental standards.

To this end, we have been consulting with industry and have refreshed the question bank and introduced new elements to the test structure. There are now behavioural case studies based on the core principles of the industry's free film *Setting out*. The film explores what employers and sites must do for workers, and how workers need to behave and react to constantly changing site conditions.

A dedicated set of respiratory risk questions have been created to help improve awareness and to promote the necessary controls. It is widely recognised that management of health risks by industry has not improved at the same rate as improvements in the management of safety risks.

Managing the environment goes hand in hand with health and safety. Therefore new questions have been introduced to help everyone understand some basic environmental responsibilities.

CITB is committed to continual improvement of delivery methods for specialist training and qualifications. We are constantly working on behalf of the industry, with the Government, Environment Agency and the HSE to drive forward improved standards.

If we all work together to support a fully competent and qualified workforce, we can have a huge impact on our industry – making it a safer and healthier place to work.

James Wates CBE

Chairman
CITB

About the test

The CITB *Health, safety and environment test* helps raise standards across the industry. It ensures that workers meet a minimum level of health, safety and environmental awareness before going on site. It should be used as a stepping stone, encouraging employers and their workforce to go on and develop their knowledge even further.

The test has been running for over 10 years and it underwent significant improvements in early 2012. These improvements included an enhanced test and a new delivery infrastructure.

All tests last for 45 minutes and have 50 multiple-choice questions including:

- ☑ 12 behavioural case study questions about how you should behave on a construction site to stay healthy and safe, followed by
- ☑ 38 knowledge questions to check your knowledge of health, safety and environment issues.

Different tests have been developed to meet the demands of different trades and professions. You will need to make sure that you take the right one.

The following tests are available:

- ☑ operative test
- ☑ specialist tests
- ☑ managers and professionals test.

What has changed in the new Health, safety and environment test?

New content has been introduced including:

- ☑ behavioural case studies
- ☑ knowledge questions on respiratory risks
- ☑ knowledge questions on environment
- ☑ a new specialist test for tunnelling
- ☑ a new structure for question selection.

What specialist tests are there?

A specialist test includes six questions from the appropriate specialist knowledge question bank (presented in Section F of this book), plus a selection of 32 core knowledge questions (presented in Sections A-E of this book).

There are currently 12 different specialist tests available including supervisory; demolition; plumbing or gas; highway works; specialist work at height; lifts and escalators; tunnelling; heating, ventilation, air conditioning and refrigeration (HVACR).

What are the behavioural case study questions?

The behavioural case study questions are designed to test how you respond to health and safety situations on a construction site. They are based on the principles established in the film *Setting out – what you should expect from a site and what they expect from you.*

Further information on the film and the transcript is provided at the back of this book.

Every test includes three case studies, each of which has four linked multiple-choice questions. These progress through a fictional situation faced by an individual working in the construction industry.

What are the knowledge questions?

The knowledge questions cover 16 core areas (presented in Sections A-E of this book) that are included in all the tests. These questions are very factual. For example they will ask you to identify fire extinguishers and signs. There is an additional knowledge question bank for each specialist test.

You will not need a detailed knowledge of the exact content or working of any regulations. However, you will need to show that you know what is required of you, the things you must do (or not do), and what to do in certain circumstances (for example, discovering an accident).

Many of the questions refer to the duties of employers. In law, the self-employed can have the same legal responsibilities as employers. To keep the questions as brief as possible, the content only refers to the duties of employers but the questions apply to both.

The questions are based on British legislation. However, Northern Ireland legislation differs from that in the rest of the UK. For practical reasons, all candidates (including those in Northern Ireland) will be tested on questions using legislation relevant to the rest of the UK only.

What is a multiple-choice question?

The test is made up of multiple-choice questions. This means you will need to select the answer you think is correct from a set of possible answers.

Most questions will ask you to select a single answer. However some of the knowledge questions will ask you to select multiple answers. This will always be clearly stated in the question.

Who writes the questions?

The question bank is developed by CITB with industry-recognised organisations which sit on or support the Health, safety and environment test question sub-committee. A full list of those parties that support the test is set out in the acknowledgements at the back of the book.

Will the questions change?

Health, safety and environment legislation, regulations and best practice will change from time to time, but CITB makes every effort to keep the test and the revision material up to date.

- You will not be tested on questions that are deemed to no longer be appropriate.

- You will be tested on knowledge questions presented in the most up to date edition of the book. To revise effectively for the test you should use the latest edition. You can check which edition of the book you have at www.citb.co.uk/hsandetestupdate or phone 0344 994 4488.

Preparing for a test

There are a number of ways you can prepare for your test.

	Operatives	Specialists	Supervisors	Managers
✪ Watch *Setting out*	Free to view at *www.citb.co.uk/settingout*			
◻ Read the question and answer books	*HS&E test for operatives and specialists* (GT 100)			*HS&E test for managers and professionals –* (GT 200)
📀 Use the digital products	*HS&E test for operatives and specialists* – *DVD* (GT 100 DVD) – *Download* (GT 100 DL) – *Apps*			*HS&E test for managers and professionals* – *DVD* (GT 200 DVD) – *Download* (GT 200 DL) – *Apps*
◻ Read supporting knowledge material	*Safe start* (GE 707)	*Safe start* (GE 707) plus sector recommended supporting material	*Site supervision simplified* (GE 706)	*Construction site safety* (GE 700)
Complete an appropriate training course	Site Safety Plus – one-day Health and safety awareness course	Contact your industry body for recommendations	Site Safety Plus – two-day Site supervisor's safety training scheme	Site Safety Plus – five-day Site management safety training scheme

How can I increase my chances of success?

☑ Prepare using the recommended revision materials, working through all the knowledge questions.

☑ Watch the *Setting out* film to prepare for the behavioural case study questions.

☑ Complete a recommended training course.

☑ Book your test when you are confident with your topics and questions.

☑ Complete a timed simulated test a few days before (available on the DVD and download).

What's on the DVD and downloads?

The DVD and downloads offer an interactive package that includes:

☑ the *Setting out* film and a sample behavioural case study

☑ all the knowledge questions and answers in both book and practice formats

☑ the real test tutorial

☑ a test simulator – all the functionality of the test with the real question bank

☑ voice-overs in English and Welsh for all questions.

☑ The DVD also provides operative questions and *Setting out* with voice-overs in the following languages:

Bulgarian, Czech, French, German, Hungarian, Lithuanian, Polish, Portuguese, Punjabi, Romanian, Russian, Spanish.

Please note: British Sign Language assistance is not included on any of the revision DVDs, except within the *Setting out* film.

Where can I buy the revision material?

CITB has developed a range of revision material, including question and answer books, DVDs, downloads and a smartphone app that will help you to prepare for the test. For further information and to buy these products:

 go online at *www.citb.co.uk/hsanderevision*

 telephone 0344 994 4488
lines open Monday to Friday 8am to 8pm, and Saturday 8am to noon

 visit a good bookshop, either in the high street or online for books and DVDs. Visit iTunes or Google Play for smartphone apps.

Where can I find more information on your supporting publications?

CITB has developed a range of publications that present a detailed and comprehensive guide to the full range of topics covered in the test. They can be used to build awareness and understanding of the issues surrounding health and safety on a construction site. They provide the context for the questions that are asked.

The range of products includes: *Safe start* (GE 707), *Site supervision simplified* (GE 706) and *Construction site safety* (GE 700). For further information and to buy these products:

 go online at *www.citb.co.uk/publications*

telephone 0344 994 4122
lines open Monday to Friday 8.30am to 5.30pm.

What is the Site Safety Plus Scheme?

CITB's Site Safety Plus Scheme is a comprehensive health and safety training programme designed to provide the building, civil engineering and allied industries with a range of courses for individuals seeking to develop their skill set in this area.

They are designed to give everyone from operative to senior manager the skill set they need to progress through the industry. From a one-day Health and safety awareness course to the five-day Site management safety training scheme (SMSTS), these courses will ensure that everyone benefits from the best possible training.

For further information:

 go online at *www.citb.co.uk/sitesafetyplus.*

Booking a test

The easiest way to book your test is either online or by telephone. You will be given the date and time of your test immediately and offered the opportunity to buy revision material, such as books, DVDs, downloads and apps. You should be able to book a test at your preferred location within two weeks.

To book your test:

 go online at *www.citb.co.uk/hsandetest*

 telephone 0344 994 4488
lines open Monday to Friday 8am to 8pm, and Saturday 8am to noon.

 post in an application form (application forms are available from the website and the telephone number listed above).

When booking your test you will be able to choose whether to receive confirmation by email or by letter. It is important that you check the details (including the type of test, the location, the date and time) and follow any instructions it gives regarding the test.

You can also choose to receive an SMS text message or email reminder 24 hours before your test.

For those instances where a test is required at short notice it may be possible to turn up at a centre and take a test on the day (subject to available spaces). It's strongly advised that you do not rely on this option.

What information do I need to book a test?

To book a test you should have the following information to hand:

- which test you need to take
- whether you require any special assistance
- your chosen method of payment (debit or credit card details)
- your address details
- your CITB registration number (you will have one of these if you have taken the test before, or applied for certain card schemes including CSCS, CPCS, CISRS, etc.).

Where can I take a test?

To sit a *Health, safety and environment test* you will need to visit a CITB-approved test centre. There are three different types of centre:

- a fixed test centre operated by our delivery provider.
  To find your nearest test centre visit *www.citb.co.uk/hsandetest*. These test centres will be able to offer tests from Monday to Saturday, but local opening times will vary. Normal opening hours are Monday to Friday 8am to 8pm and Saturdays 8am to noon
- an independent Internet Test Centre (for example as operated by a college, training provider or commercial organisation). Please contact these test centres directly for further information and to book a test

- a corporate testing unit which can be established at a suitable venue for a group of candidates. For further information on this service there is a dedicated booking line 0344 994 4492.

Is there any special assistance available when taking the test?

Voice-over assistance

All tests can be booked with English or Welsh voice-overs.

Foreign language assistance

- The operatives test can be booked with voice-overs in the following languages: Bulgarian, Czech, French, German, Hungarian, Lithuanian, Polish, Portuguese, Punjabi, Romanian, Russian, Spanish.

  An interpreter can be requested if assistance is required in other languages.

- The specialist tests can be booked with an interpreter but no pre-recorded voice-overs arc available.

- The managers and professionals test does not allow foreign language assistance because a basic command of English or Welsh is required in order to sit the test.

Sign language assistance

The operatives test can be booked with British Sign Language on screen. If you need assistance in the other tests a signer can be provided.

Further assistance

If you need any other special assistance (such as a reader, signer, interpreter, or extra time) this can be provided but you will need to book through a dedicated booking line 0344 994 4491.

What services are there for Welsh speakers?

- All tests can be booked with Welsh voice-overs.
- All revision DVDs include Welsh voice-overs.
- There is a dedicated Welsh booking line 0344 994 4490.

How do I cancel or postpone my test?

To cancel or reschedule your test you should go online or call the booking number at least 72 hours (three working days) before your test, otherwise you will lose your fee.

What if I do not receive a confirmation email/letter?

If you do not receive a confirmation email or letter within the time specified please call the booking line to check your booking has been made.

We cannot take responsibility for postal delays. If you miss your test event, you will unfortunately forfeit your fee.

Taking a test

Before the test

On the day of the test you will need to:

- allow plenty of time to get to the test centre and arrive at least 15 minutes before the start of the test
- take your confirmation email or letter
- take proof of identity that bears your photo and your signature (such as driving licence card or passport – please visit *www.citb.co.uk/hsandetest* for full list of acceptable documentation).

On arrival at the test centre, staff will check your documents to ensure you are booked onto the correct test. If you do not have all the relevant documents you will not be able to sit your test and you will lose your fee.

During the test

The tests are all delivered on a computer screen. However, you do not need to be familiar with computers and the test does not involve any typing. All you need to do is click on the relevant answer boxes, using either a mouse or by touching the screen.

Before the test begins you can choose to work through a tutorial. It explains how the test works and lets you try out the buttons and functions that you will use while taking your test.

The test will contain 50 multiple-choice questions which you will need to complete in 45 minutes.

There will be information displayed on the screen which shows you how far you are through the test and how much time you have remaining.

After the test

At the end of the test there is an optional survey that gives you the chance to provide feedback on the test process.

You will be provided with a printed score report after you have left the test room. This will tell you whether you have passed or failed your test, and give feedback on areas where further learning and revision is recommended.

What do I do if I fail?

If you fail your test, your score report will provide feedback on areas where you got questions wrong.

It is strongly recommended that you revise these areas thoroughly before re-booking. You will have to wait at least 72 hours before you can take the test again.

What do I do if I pass?

A *Health, safety and environment test* pass is often a necessary requirement when applying to join a construction industry card scheme. Different schemes exist in different trades and professions. Membership of a relevant scheme helps you prove that you can do your job, and that you can do it safely. Access to construction sites may require a relevant scheme card.

Once you have passed your test, you should, if you have not done so already, consider applying to join the relevant card scheme. However please be aware that you may need to complete further training, assessment and/or testing to meet their specific entry requirements.

To find out more about many of the recognised schemes:

 go online at *www.citb.co.uk/cardschemes.*

Further scheme contact details		telephone	go online
General	Construction Skills Certification Scheme (CSCS)	0344 994 4777	www.cscs.uk.com
	Northern Ireland: Construction Skills Register (CSR)	028 9087 7150	www.cefni.co.uk
Plant operatives	Construction Plant Competence Scheme (CPCS)	0844 815 7274	www.citb.co.uk/cpcs
Demolition operatives	Certificate of Competence of Demolition Operatives (CCDO)	0844 826 8385	www.ndtg.org
Scaffolders	Construction Industry Scaffolders Record Scheme (CISRS)	0844 815 7223	www.cisrs.org.uk
HVACR	Engineering Services SKILLcard (ESS)	01768 860 406	www.skillcard.org.uk
Plumbers	Joint Industry Board for Plumbing and Mechanical Engineering Services (JIB-PMES)	01480 476 925	www.jib-pmes.org
	Scottish and Northern Ireland Joint Industry Board (SNIJIB)	0131 556 0600	www.snijib.org
Electricians	Electrotechnical Certification Scheme (ECS)	In England call 03333 218 230 In Scotland call 0131 445 5577	www.ecscard.org.uk

A

Legal and management

Contents

General responsibilities

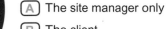
1.1

Who is responsible for reporting any unsafe conditions on site?

- A The site manager only
- B The client
- C Health and Safety Executive (HSE) inspectors
- D Everyone on site

1.2

During site induction you do not understand something the presenter says. What should you do?

- A Attend another site induction
- B Ask the presenter to explain the point again
- C Guess what the presenter was trying to tell you
- D Wait until the end then ask someone else to explain

1.3

Now that work on site is moving forward, the safety rules given in your site induction seem out of date. What should you do?

- A Do nothing, you are not responsible for safety on site
- B Speak to your supervisor about your concerns
- C Speak to your workmates to see if they have any new rules
- D Make up your own safety rules to suit the changing conditions

1.4

What is the most important reason for keeping your work area clean and tidy?

- A To prevent slips, trips and falls
- B So that you don't have a big clean-up at the end of the week
- C So that waste skips can be emptied more often
- D To recycle waste and help the environment

1.5

The work of another contractor is affecting your safety. You should stop work and:

- A go home
- B speak to your supervisor
- C speak to the contractor's supervisor
- D speak to the contractor who is doing the job

1.6

You are using some equipment. It has just been given a prohibition notice. What does this mean?

- A You must not use it unless your supervisor is present
- B You must not use it until it is made safe
- C You can use it as long as you take more care
- D Only supervisors can use it

Answers: 1.1 = D   1.2 = B   1.3 = B   1.4 = A   1.5 = B   1.6 = B

A
01

1.7

The whole site has been issued with a prohibition notice. What does this mean to you?

- A) You can carry on working because it was nothing to do with you
- B) You are not allowed to use any power tools
- C) You can finish what you are doing and then go home
- D) You must stop work because the site is unsafe

1.8

After watching you work, a Health and Safety Executive (HSE) inspector issues an improvement notice. What does this mean?

- A) You are not working fast enough
- B) You need to improve the standard of your work
- C) You are not working in a safe way
- D) All of these answers

1.9

The Law (Health and Safety at Work Act) places legal duties on:

- A) employers only
- B) operatives only
- C) all people at work
- D) self-employed people only

1.10

As a worker you do NOT have a legal duty to:

- A) use all equipment safely and as instructed
- B) write your own risk assessments
- C) speak to your supervisor if you are worried about safety on site
- D) report any equipment that is damaged or defective

1.11

Why is the Health and Safety at Work Act important to you? Give TWO answers.

- A) It tells you which parts of the site are dangerous
- B) It must be learned before starting work
- C) It puts legal duties on your employer to provide a safe place to work
- D) It tells you how to do your job
- E) It puts legal duties on you as a worker

1.12

Which of these is NOT your legal duty as a worker?

- A) To look after your own health and safety
- B) To look after the health and safety of anyone else who might be affected by your work
- C) To write your own risk assessments
- D) Not to interfere with anything provided for health and safety

Answers: 1.7 = D 1.8 = C 1.9 = C 1.10 = B 1.11 = C, E 1.12 = C

A
01

1.13

Who is responsible for managing health and safety on construction sites?

A The company safety officer

B The Health and Safety Executive (HSE)

C The client

D The site manager

1.14

A risk assessment identifies:

A how to report accidents

B the site working hours

C the hazards and safe way of doing the job

D where the first-aid box is kept

1.15

You will often hear the word hazard mentioned. What does it mean?

A Anything at work that could harm you

B The site accident rate

C A type of barrier or machine guard

D All of these answers

1.16

You are about to start a job. How will you know if it needs a permit to work?

A Other workers will tell you

B The Health and Safety Executive (HSE) will tell you

C You will not be allowed to start until the permit to work has been issued

D You don't need to know, as permits to work only affect managers

1.17

Which THREE of the following should be included in a method statement?

A The materials, tools and equipment needed

B The people involved and the level of competency and training required

C The directions to the site

D The order and correct way the work should be done

E The risks you can take

1.18

You find that you cannot do a job as the method statement says you should. What do you do?

A Make up your own way of doing the job

B Do not start work until you have talked with your supervisor

C Ask your workmates how they think you should do the job

D Contact the Health and Safety Executive (HSE)

1.19

A permit to work allows:

A the emergency services to come on to the site after an accident

B certain jobs to be carried out safely under controlled conditions,

C Health and Safety Executive (HSE) inspectors to visit the site

D untrained people to work without supervision

1.20

How would you expect to find out about site health and safety rules when you first arrive on site?

A During site induction.

B In a letter sent to your home

C By reading your employer's health and safety policy

D By asking others on the site

1.21

It is your employer's legal responsibility to discuss matters of health and safety with you because:

A it will mean that you will never have to attend any other health and safety training

B having done so, your employer will not have any legal responsibility for your health and safety

C they must inform you of things that will protect your health and safety.

D you do not have any responsibilities for health and safety

1.22

What is a toolbox talk?

A A short training session on a particular safety topic,

B A talk that tells you where to buy tools

C Your first training session when you arrive on site

D A sales talk given by a tool supplier

A
01

1.23

Who should attend a site induction?

A Cleaners

B Architects

C Construction-related workers

D Everyone going on to the site,

1.24

During a site induction, which of the following TWO topics should be covered?

A The site rules,

B Where the cheapest car park is

C Holiday dates

D The site emergency procedures,

E Information on local amenities

Answers: 1.19 = B   1.20 = A   1.21 = C   1.22 = A   1.23 = D   1.24 = A, D

15

1.25

A near miss is an incident where:

A you were just too late to see what happened

B someone could have been injured

C someone was injured and nearly had to go to hospital

D someone was injured and nearly had to take time off work

1.26

You can help prevent accidents by:

A reporting unsafe working conditions

B becoming a first aider

C knowing where the first-aid kit is kept

D knowing how to get help quickly

1.27

What is the MAIN reason for attending a site induction?

A You will get to know other new starters

B Risk assessments will be handed out

C Site health and safety rules and site hazards will be explained

D Permits to work will be handed out

1.28

What are TWO possible consequences for employers if they do NOT take measures to prevent accidents and ill health at work?

A They could be fined or imprisoned

B They could win more work

C They will lose time and money due to the cost of any accident or ill health

D Their workforce will be happier

E They will get a good reputation

1.29

What are TWO possible consequences if accidents and ill health are NOT prevented at work?

A You will have to work harder

B You may suffer a personal injury

C You may get a bonus

D You may not be able to work, which could impact on your personal income and family life

E You will live longer

Answers: 1.25 = B   1.26 = A   1.27 = C   1.28 = A, C   1.29 = B, D

2.1

Which TWO of the following will help you find out about the site emergency procedures and emergency telephone numbers?

- A) Guidance from the Health and Safety Executive (HSE)
- B) Reading the site noticeboards
- C) Guidance from your local Job Centre Plus
- D) Attending the site induction
- E) Looking in the telephone directory

2.2

In an emergency you should:

- A) leave site
- B) phone home
- C) follow the site emergency procedure
- D) phone the Health and Safety Executive (HSE)

2.3

In an emergency an 'assembly point' is the:

- A) site manager's office
- B) place of the incident or accident
- C) welfare facilities
- D) specified place to gather after an evacuation

2.4

A scaffold has collapsed and you saw it happen. When you are asked about the accident, you should say:

- A) nothing, you are not a scaffold expert
- B) as little as possible because you don't want to get people into trouble
- C) exactly what you saw
- D) who you think is to blame and how they should be punished

A 02

2.5

What is the MAIN objective of carrying out an accident investigation?

- A) To find out who is at fault
- B) To find out the causes in order to prevent it happening again
- C) To find out the cost of any damage that occurred
- D) To record what injuries were sustained

2.6

You have witnessed a serious accident on your site and are to be interviewed by a Health and Safety Executive (HSE) inspector. Should you:

- A) ask your supervisor what you should say to the inspector
- B) not tell the inspector anything
- C) co-operate and tell the inspector exactly what you saw
- D) tell the inspector what your workmates have told you

Answers: 2.1 = B, D   2.2 = C   2.3 = D   2.4 = C   2.5 = B   2.6 = C

A
02

2.7

When must you record an accident in the accident book?

- [A] If you are injured in any way
- [B] Only if you have to be off work
- [C] Only if you have suffered a broken bone
- [D] Only if you have to go to hospital

2.8

If someone is injured at work, who should record it in the accident book?

- [A] The site manager and no-one else
- [B] The injured person or someone acting for them
- [C] The first aider and no-one else
- [D] Someone from the Health and Safety Executive (HSE)

2.9

Which of these does NOT have to be recorded in the accident book?

- [A] Your national insurance number
- [B] The date and time of your accident
- [C] Details of your injury
- [D] Your home address

2.10

Which type of accidents should be recorded in the accident book?

- [A] Only specified injuries, such as a broken arm or death of a person
- [B] Only injuries requiring hospital treatment
- [C] All accidents causing any injury
- [D] Only accidents where the injured person has to stop work

2.11

When must an accident be recorded in the site's accident book?

- [A] Only when an accident causes injury to a worker while at work
- [B] Only when a person is injured and will be off work for more than three days
- [C] Only when an accident causes damage to plant or equipment
- [D] Only when a person breaks a major bone or is concussed

2.12

An entry must be made in the accident book when:

- [A] an accident causes personal injury to any worker
- [B] the person has been off sick for three days
- [C] the severity of the accident may result in a compensation claim
- [D] management thinks it appropriate

2.13

Which of the following items is NOT recorded in an accident book?

A Your national insurance number

B The date and time of the accident

C The injuries sustained

D Your home address

2.14

You suffer an injury at work and the details are recorded into the accident book. What MUST happen to this accident record?

A It must be sent to the employer's insurance company at the end of the job

B It must be kept in a place where anyone can read it

C It must be treated as confidential under the Data Protection Act and kept for at least three years

D It must be destroyed at the end of the job

2.15

Accidents causing any injury should always be recorded in the:

A main contractor's diary

B accident report book

C site engineer's day book

D sub-contractor's diary

2.16

Why is it important to report all accidents?

A It might stop them happening again

B Some types of accident have to be reported to the Health and Safety Executive (HSE)

C Details have to be entered in the accident book

D All of these answers

2.17

Why is it important to report near miss incidents on site?

A Because it is the law for all 'near miss' incidents

B To find someone to blame

C It is a requirement of the CDM Regulations

D To learn from them and stop them happening again

2.18

If you have a minor accident, who should report it?

A Anyone who saw the accident happen

B A sub-contractor

C You

D The Health and Safety Executive (HSE)

A
02

A
02

2.19

If your doctor says that you have Weil's disease (leptospirosis), contracted when on site, why do you need to tell your employer?

- A Your employer has to warn your colleagues not to go anywhere near you
- B Your employer will have to report it to the Health and Safety Executive (HSE)
- C Your work colleagues might catch it from you
- D The site on which you contracted it will have to be closed down

2.20

While working on site you get a small cut on one of your fingers. What should you do?

- A Report it at the end of the day or the end of the shift
- B Wash it, and if it is not a problem carrying on working
- C Clean it up and tell your supervisor about it later
- D Report it and get first aid if necessary

2.21

You receive an injury from an accident at work. When should you report it?

- A At the end of the day, before you go home
- B Only if you had to take time off work
- C Immediately, or as soon as possible afterwards
- D The next day before you start work

2.22

You have suffered an injury caused by an accident at work and as a result you are absent for more than seven days. Which TWO of the following actions MUST be taken?

- A The accident must be recorded in the site accident book
- B The emergency services are called to assess the circumstances of the accident
- C The local hospital and the benefits office must be informed
- D Your employer should inform the Health and Safety Executive (HSE)
- E You must pay for any first-aid equipment used to treat your injury

2.23

Why should you report an accident?

- A It helps the site find out who caused it
- B It is a legal requirement
- C So that the site manager can see who is to blame
- D So that your company will be held responsible

2.24

Who must you report a serious accident to?

- A Site security
- B The police service
- C Your employer
- D The ambulance service

Answers: 2.19 = B   2.20 = D   2.21 = C   2.22 = A, D   2.23 = B   2.24 = C

2.25

You are involved in an incident on site that was dangerous but no-one was injured. Who must you report this 'dangerous occurrence' to?

A Your site supervisor or the site manager

B The client for the project

C The rest of the workforce

D The first aider

2.26

You have witnessed a serious accident on your site. Should you:

A say nothing to anyone in case you get someone into trouble

B ask your workmates what they think you should do

C telephone the local hospital

D tell your supervisor that you saw what happened

2.27

Your doctor tells you that you have hand-arm vibration syndrome possibly caused through work. What should you do?

A Tell no-one as it's embarrassing

B Inform your site supervisor or employer

C Just inform your workmates

D Tell no-one as this is not reportable

A
02

A
03

3.1

You will find out about emergency assembly points from:

A a risk assessment

B a method statement

C the site induction

D the permit to work

3.2

How should you be informed about what to do in an emergency? Give TWO answers.

A From the site induction

B Look in the health and safety file

C Ask the Health and Safety Executive (HSE)

D Ask the local hospital

E From the site noticeboards

3.3

A first-aid box should NOT contain:

A bandages

B plasters

C safety pins

D over the counter medicines such as aspirin or painkillers

3.4

The first-aid box on site is always empty. What should you do?

A Bring your own first-aid supplies into work

B Find out who is taking all the first-aid supplies

C Find out who looks after the first-aid box and let them know

D Ignore the problem, it is always the same

3.5

Does your employer have to provide a first-aid box?

A Yes, every site must have one

B Only if more than 50 people work on site

C Only if more than 25 people work on site

D No, there is no legal duty to provide one

3.6

When would you expect eyewash bottles to be provided?

A Only on demolition sites where asbestos has to be removed

B Only on sites where refurbishment is being carried out

C On all sites where people could get something in their eyes

D On all sites where showers are needed

3.7

If you want to be a first aider, you should:

(A) watch a first aider treating people then try it yourself

(B) ask if you can do a first-aider's course

(C) buy a book on first aid and start treating people

(D) speak to your doctor about it

3.8

What is the first thing you should do if you find an injured person?

(A) Tell your supervisor

(B) Check that you are not in any danger before you check the injured person

(C) Move the injured person to a safe place

(D) Ask the injured person what happened

3.9

If someone falls and is knocked unconscious, you should first:

(A) turn them over so they are lying on their back

(B) send for medical help

(C) slap their face to wake them up

(D) give mouth-to-mouth resuscitation

3.10

Someone has fallen from height and has no feeling in their legs. You should:

(A) roll them onto their back and keep their legs straight

(B) roll them onto their side and bend their legs

(C) ensure they stay still and don't move them until medical help arrives

(D) raise their legs to see if any feeling comes back

3.11

Someone working in a deep manhole has collapsed. What is the first thing you should do?

(A) Get someone to lower you into the manhole on a rope

(B) Climb into the manhole and give mouth-to-mouth resuscitation

(C) Go and tell your supervisor

(D) Shout and raise the alarm as a trained rescue team will be needed

3.12

If there is an emergency while you are on site you should first:

(A) leave the site and go home

(B) phone home

(C) follow the site emergency procedure

(D) phone the Health and Safety Executive (HSE)

A
03

Answers: 3.7 = B   3.8 = B   3.9 = B   3.10 = C   3.11 = D   3.12 = C

A
03

3.13

If someone is in contact with a live cable the best thing you can do is:

- A) phone the electricity company
- B) dial 999 and ask for an ambulance
- C) switch off the power and call for help
- D) pull them away from the cable

3.14

 What does this sign mean?

- A) Safety glasses cleaning station
- B) Emergency eyewash station
- C) Warning, risk of splashing
- D) Wear eye protection

3.15

 What does this sign mean?

- A) First aid
- B) Safe to cross
- C) No waiting
- D) Medicine box

3.16

If you cut your finger and it won't stop bleeding, you should:

- A) wrap something around it and carry on working
- B) tell your workmates
- C) wash it clean then carry on working
- D) find a first aider or get other medical help

3.17

What is the one thing a first aider CANNOT do for you?

- A) Give mouth-to-mouth resuscitation
- B) Stop any bleeding
- C) Give you medicines without authorisation
- D) Treat you if you are unconscious

3.18

If you think someone has a broken leg you should:

- A) lie them on their side in the recovery position
- B) use your belt to strap their legs together
- C) send for the first aider or get other help
- D) lie them on their back

3.19

If someone gets some grit in their eye, the best thing you can do is:

A) hold the eye open and wipe it with clean tissue paper

B) ask them to rub the eye until it starts to water

C) tell them to blink a couple of times

D) hold the eye open and flush it with sterilised water or eyewash

3.20

Someone gets a large splinter in their hand. It is deep under the skin and it hurts. What should you do?

A) Use something sharp to dig it out

B) Make sure they get first aid

C) Tell them to ignore it and let the splinter come out on its own

D) Try to squeeze out the splinter with your thumbs

3.21

Someone collapses with stomach pain and there is no first aider on site. What should you do first?

A) Get them to sit down

B) Get someone to call the emergency services

C) Get them to lie down in the recovery position

D) Give them some painkillers

3.22

Someone has got a nail in their foot. You are not a first aider. You must not pull out the nail because:

A) you will let air and bacteria get into the wound

B) the nail is helping to reduce the bleeding

C) it will prove that the casualty was not wearing safety boots

D) the nail is helping to keep their boot on

3.23

If someone burns their hand the best thing you can do is:

A) put the hand into cold water or under a cold running tap

B) tell them to carry on working to exercise the hand

C) rub barrier cream or Vaseline into the burn

D) wrap your handkerchief around the burn

A
03

B

Health and welfare

Contents

B
04

4.1

It is your first day on site. You find that there is nowhere to wash your hands. What should you do?

A) Wait until you get home, then wash them

B) Go to a local café or pub and use the washbasin in their toilet

C) Speak to your supervisor about the problem

D) Bring your own bottle of water the next day

4.2

Look at these statements about illegal drugs in the workplace. Which one is true in relation to site work?

A) Users of illegal drugs are a danger to everyone on site

B) People who take illegal drugs work better and faster

C) People who take illegal drugs take fewer days off work

D) Taking illegal drugs is a personal choice so other people shouldn't worry about it

4.3

Your doctor has given you some medication. Which of these questions is the most important?

A) Will it make me drowsy or unsafe to work?

B) Will I work more slowly?

C) Will my supervisor find out?

D) Will I oversleep and be late for work?

4.4

Someone goes to the pub at lunchtime and has a couple of pints of beer. What should they do next?

A) Drink plenty of strong coffee then go back to work

B) Stay away from the site for the rest of the day

C) Stay away for an hour and then go back to work

D) Eat something, wait 30 minutes and then go back to work

4.5

You should only clean very dirty hands with:

A) soap and water

B) thinners

C) white spirit

D) paraffin

4.6

If you get a hazardous substance on your hands, it can pass from your hands to your mouth when you eat. Give TWO ways to stop this.

A) Wear protective gloves while you are working

B) Wash your hands before eating

C) Put barrier cream on your hands before eating

D) Wear protective gloves then turn them inside-out before eating

E) Wash your work gloves then put them on again before eating

 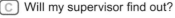

4.7

You can catch an infection called tetanus from contaminated land or water. How does it get into your body?

- A. Through your nose when you breathe
- B. Through an open cut in your skin
- C. Through your mouth when you eat or drink
- D. It doesn't, it only infects animals and not people

4.8

You should not use white spirit or other solvents to clean your hands because they:

- A. strip the protective oils from the skin
- B. remove the top layer of skin
- C. block the pores of the skin
- D. carry harmful bacteria that attack the skin

4.9

Direct sunlight on bare skin can cause:

- A. dermatitis
- B. rickets
- C. acne
- D. skin cancer

4.10

You should not just rely on barrier cream to protect your skin from harmful substances because:

- A. it costs too much to use every day
- B. many harmful substances go straight through it
- C. it is difficult to wash off
- D. it can irritate your skin

4.11

If you use skin barrier cream you should apply it:

- A. before you start work
- B. when you finish work
- C. as part of first-aid treatment
- D. when you can't find your gloves

4.12

Exposure to engine oil and other mineral oils can cause:

- A. skin problems
- B. heart disease
- C. breathing problems
- D. vibration white finger

B
04

B
04

4.13

You can get occupational dermatitis from:

A) hand-arm vibration

B) another person with dermatitis

C) some types of strong chemical

D) sunlight

4.14

Which of the following species of animal is the most likely carrier of Weil's disease (leptospirosis) on construction sites?

A) Rabbits

B) Rats

C) Squirrels

D) Mice

4.15

To help keep rats away, everyone on site should:

A) buy rat traps and put them around the site

B) ask the local authority to put down rat poison

C) throw food scraps over the fence or hoarding

D) only put food and drink rubbish in bins provided

4.16

You are more likely to catch Weil's disease (leptospirosis) if you:

A) work near wet ground, waterways or sewers

B) work near air-conditioning units

C) fix showers or baths

D) drink water from a standpipe

4.17

The early signs of Weil's disease (leptospirosis) can be easily confused with:

A) dermatitis

B) diabetes

C) hay fever

D) influenza (flu)

4.18

What sort of rest area should your employer provide on site?

A) A covered area

B) A covered area and some chairs

C) A covered area, tables and chairs, and something to heat water and food

D) Nothing, employers don't have to provide rest areas

4.19

What is the legal minimum that should be provided on site for washing your hands?

A Nothing, there is no need to provide washing facilities

B A bowl, kettle and towel

C A cold water standpipe and paper towels

D Hot and cold water (or warm water), soap and a way to dry your hands

B
04

4.20

The toilets on your site are always dirty or do not flush. What should you do?

A Try not to use the toilets while you are at work

B Tell the person in charge of the site about the problem

C Try to fix the fault yourself

D See if you can use the toilets in a nearby café or pub

Answers: 4.19 = D   4.20 = B

B
05

5.1

When do you need to wear eye protection?

A On very bright, sunny days

B If there is a risk of eye injury and if it is the site rules

C When your employer can afford it

D Only if you work with chemicals

5.2

If there is a risk of materials flying into your eyes, you should wear:

A tinted welding goggles

B laser safety glasses

C chemical-resistant goggles

D impact-resistant goggles

5.3

What type of eye protection do you need to wear if you are using a cartridge-operated tool or compressed gas tool (nail gun)?

A Light eye protection or safety glasses

B Normal prescription glasses or sunglasses

C Impact-rated goggles

D None – they aren't needed as there is a minimal risk of injury

5.4

You need to use a grinder, cut-off saw, cartridge tool or nail gun. What type of eye protection will you need?

A Impact-resistant goggles or full face shield

B Welding goggles

C Reading glasses or sunglasses

D Light eye protection (safety glasses)

5.5

Safety footwear with a protective mid-sole will protect you against:

A twisting your ankle

B chemicals burning your feet

C nails or sharp objects puncturing the underside of your foot if you stand on them

D getting blisters

5.6

When should you wear safety footwear on site?

A Only when working at ground level or outside

B Until the site starts to look finished

C All the time

D When you are working all day on site and not just visiting

5.7

You need to handle a hazardous substance. You should wear the correct gloves to help stop:

(A) skin disease

(B) vibration white finger

(C) Raynaud's syndrome

(D) arthritis

5.8

Do all types of glove protect hands against chemicals?

(A) Yes, all gloves are made to the same standard

(B) Only if you put barrier cream on your hands as well

(C) No, different types of glove protect against different types of hazard

(D) Only if you cover the gloves with barrier cream

5.9

Look at these statements about using power tools while wearing gloves. Which one is FALSE?

(A) Anti-vibration gloves will protect you against the effects of vibration

(B) Gloves will keep your hands warm and dry, which reduces the effects of vibration

(C) Gloves provide a better grip so you don't need to grip too tightly, which reduces the effects of vibration

(D) Gloves will protect you from cuts and abrasions

5.10

You need to wear a full body harness. You have never used one before. What should you do?

(A) Ask for expert advice and training

(B) Ask someone already wearing a harness to show you what to do

(C) Try to work it out for yourself

(D) Read the instruction book

B
05

5.11

You must wear head protection on site at all times unless you are:

(A) self-employed

(B) working alone

(C) in a safe area, like the site office or canteen

(D) working in very hot weather

5.12

To get the maximum protection from your safety helmet you should wear it:

(A) back to front

(B) pushed back on your head

(C) square on your head

(D) pulled forward

5.13

Look at these statements about wearing a safety helmet in hot weather. Which one is true?

A You can drill holes in it to keep your head cool

B You can wear it back-to-front if it is more comfortable that way

C You must take it off during the hottest part of the day

D You must wear it at all times and in the right way

5.14

If you drop your safety helmet from height onto a hard surface, you should:

A have any cracks repaired then carry on wearing it

B make sure there are no cracks then carry on wearing it

C work without a safety helmet until you can get a new one

D stop work and get a new safety helmet

5.15

You have been given disposable earplugs to use, but they keep falling out. What should you do?

A Throw them away and work without them

B Stop work until you get more suitable ones and are shown how to fit them

C Put two earplugs in each ear so they stay in place

D Put rolled-up tissue paper in each ear

5.16

Your employer must supply you with personal protective equipment (PPE):

A twice a year

B if you pay for it

C if it is in the contract

D if you need to be protected

5.17

Do you have to pay for any personal protective equipment (PPE) you need?

A Yes, you must pay for all of it

B Only if you need to replace lost or damaged PPE

C Yes, but you only have to pay half the cost

D No, your employer must pay for it

5.18

Who should provide you with any personal protective equipment (PPE) you need, including the means to maintain it?

A Your employer

B You must buy your own

C Anyone on site with a budget

D No-one has a duty to provide it

5.19

If your personal protective equipment (PPE) gets damaged you should:

- A throw it away and work without it
- B stop what you are doing until it is replaced
- C carry on wearing it but work more quickly
- D try to repair it

5.20

You have to work outdoors in bad weather. Your employer should supply you with correctly fitting waterproof clothing because:

- A it will have the company name and logo on it
- B you need protecting from the weather and are less likely to get muscle strains if you are warm and dry
- C you are less likely to catch Weil's disease (leptospirosis) if you are warm and dry
- D your supervisor will be able to see you more clearly in the rain

5.21

Look at these statements about personal protective equipment (PPE). Which one is NOT true?

- A You must pay for any damage or loss
- B You must store it correctly when you are not using it
- C You must report any damage or loss to your supervisor
- D You must use it as instructed

5.22

You are about to start a job. How will you know if you need any extra personal protective equipment (PPE)?

- A By looking at your employer's health and safety policy
- B You will just be expected to know
- C From the risk assessment or method statement
- D A letter will be sent to your home

B
05

Answers: 5.19 = B   5.20 = B   5.21 = A   5.22 = C

Dust and fumes (Respiratory risks)

6.1

If using on-tool extraction to control dust from a power tool it is important to check that:

- A) the extraction unit is the correct type
- B) the extraction filters are clear and the unit is extracting dust
- C) you are using the power tool correctly
- D) all of these answers

6.2

When drilling, cutting, sanding or grinding you can breathe in high levels of harmful dust. These levels are likely to be highest when working:

- A) outside on a still day
- B) outside on a windy day
- C) in a small room
- D) in a large indoor space

6.3

You have finished your work and need to sweep up the dust created. What should you do?

- A) Dampen down the area
- B) Make sure there is plenty of ventilation
- C) Put your protective mask back on
- D) All of these answers

6.4

You are using water as part of dust control and run out. Should you:

- A) carry on as you have nearly finished
- B) stop and refill with water
- C) ask everyone to clear the area and then carry on
- D) carry on but get someone to sweep up afterwards

6.5

When using power tools it is important to:

- A) stop dust getting into the air
- B) stand downwind of any dust
- C) do the work quickly to limit dust exposure
- D) only undertake the work during damp or wet weather

6.6

You have been asked to do some work that will create dust. What should you do?

- A) You should not do the work. Dust is highly dangerous
- B) Use equipment that will eliminate or reduce the amount of dust whilst wearing correct personal protective equipment (PPE)
- C) Start work – no controls are needed as it's only dust
- D) Work for short periods at a time

6.7

When using water to keep dust down when cutting you must ensure:

A. there is as much water as possible

B. the water flow is correctly adjusted

C. somebody stands next to you and pours water from a bottle

D. water is poured onto the surface to soak it, before you start cutting

6.8

You need to use a power tool to cut or grind materials. Give TWO ways to control the dust from getting into the air.

A. Work slowly and carefully

B. Fit a dust extractor or collector to the machine

C. Wet cutting

D. Keep the area clean and tidy

E. Wear a dust mask or respirator

6.9

If you use a power tool to cut or grind materials, why must the dust be collected and not get into the air?

A. To save time and avoid having to clear up the mess

B. Most dust can be harmful if breathed in

C. The tool will go faster if the dust is collected

D. You do not need a machine guard if the dust is collected

6.10

Occupational asthma can stop you working again with certain substances. It is caused by:

A. exposure to loud noise

B. exposure to rat urine

C. skin contact with any hazardous substance

D. breathing in hazardous dust, fumes or vapours

6.11

There are many kinds of dust and fumes at work. Breathing them in over time can cause you to develop:

A. occupational lung disease

B. occupational dermatitis

C. skin cancer

D. sore throat

6.12

Exposure to which of the following may NOT result in lung disease?

A. Asbestos

B. Bird droppings

C. Strong smells

D. Silica dust

B
06

B
06

6.13

Pigeon droppings and nests, which can be hazardous to your health, are found in an area where you are required to work. You should:

A carry on with your work carefully

B stop work and seek advice

C try to catch the pigeons

D let them fly away before carrying on with your work

6.14

What is the biggest cause of LONG-TERM health issues in the construction industry?

A Loud noise

B Being struck by a vehicle

C Slipping and tripping

D Breathing in hazardous substances

6.15

You need special respiratory protective equipment (RPE) to handle a chemical. None has been provided. What should you do?

A Get on with the job but try to work quickly

B Do not start work until you have been given the correct RPE and training

C Start the work but take a break now and again

D Sniff the substance to see if it makes you feel ill

6.16

You have been given a dust mask to protect you against hazardous fumes. What should you do?

A Do not start work until you have the correct respiratory protective equipment (RPE)

B Do the job but work quickly

C Start work but take a break now and again

D Wear a second dust mask on top of the first one

6.17

Which of the following do you need to do to ensure that your mask works?

A Check it's the correct type needed

B Pass a face-fit test wearing the mask

C Check you are wearing it correctly

D All of these answers

6.18

Generally speaking how long can you use the same disposable mask for?

A Five working days

B Until it looks too dirty to wear

C One day or one shift

D 28 days

6.19

Which of these activities does NOT create silica dust, which is harmful if breathed in?

A. Sawing timber and plywood

B. Cutting kerbs, stone, paving slabs, bricks and blocks

C. Breaking up concrete floors and screeds

D. Chasing out walls and mortar joints or sweeping up rubble

6.20

When drilling, cutting, sanding or grinding what is the best way to protect your long-term health from harmful dust?

A. Use dust extraction or wet cut and wear light eye protection

B. Wear a FFP3-rated dust mask and impact goggles

C. Wear any disposable dust mask, hearing protection and impact goggles

D. Use dust extraction or wet cut, wear a FFP3-rated dust mask, hearing protection and impact goggles

6.21

The high levels of solvents in some paints and resins can cause:

A. headaches, dizziness and sickness

B. lung problems

C. effects on other parts of your body

D. all of these answers

B
06

B
07

7.1

Noise can damage your hearing. What is an early sign of this?

- A There are no early signs
- B Temporary deafness or ringing noise in your ears
- C A skin rash around the ears
- D Ear infections

7.2

After working with noisy equipment you have a 'ringing' sound in your ears. What does this mean?

- A Your hearing has been temporarily damaged
- B You have also been subjected to vibration
- C You are about to go down with the flu
- D The noise level was high but acceptable

7.3

Noise over a long time can damage your hearing. Can this damage be reversed?

- A Yes, with time
- B Yes, if you have an operation
- C No, the damage is permanent
- D Yes, if you change jobs

7.4

How can noise affect your health? Give TWO answers.

- A Headaches
- B Ear infections
- C Hearing loss
- D Waxy ears
- E Vibration white finger

7.5

You think the noise at work may have damaged your hearing. What should you do?

- A Plug your ears with cotton wool to stop any more damage
- B Nothing, the damage has already been done
- C Go off sick
- D Ask your employer or doctor to arrange a hearing test

7.6

Someone near you is using noisy equipment and you have no hearing protection. What should you do?

- A Ask them to stop what they are doing
- B Carry on with your work because it is always noisy on site
- C Leave the area until you have the correct personal protective equipment (PPE)
- D Speak to the other person's supervisor

7.7

If you wear hearing protection, it will:

A) stop you hearing all noise

B) reduce damaging noise to an acceptable level

C) repair your hearing if it is damaged

D) make you hear better

7.8

If you need to wear disposable earplugs how should you insert them so they protect your hearing from damage?

A) Only put them in when it starts getting very noisy

B) Only ever insert them half way into your ear

C) Roll them up and insert them as far as you can, while pulling the top of your ear up to open up the ear canal

D) Fold them in half and wedge them into your ear

7.9

TWO recommended ways to protect your hearing are by using:

A) rolled tissue paper

B) cotton wool pads over your ears

C) earplugs

D) soft cloth pads over your ears

E) ear defenders

7.10

As a rule of thumb noise levels may be a problem if you have to shout to be clearly heard by someone who is standing:

A) 2 m away

B) 4 m away

C) 5 m away

D) 6 m away

7.11

You need to wear ear defenders but an ear pad is missing from one of the shells. What should you do?

A) Leave them off and work without any hearing protection

B) Put them on and start working with them as they are

C) Do not work in noisy areas until they are replaced

D) Wrap your handkerchief around the shell and carry on working

7.12

If you have to work in a 'hearing protection zone', you must:

A) not make any noise

B) wear the correct hearing protection at all times

C) take hearing protection with you in case you need to use it

D) wear hearing protection if the noise gets too loud for you

B
07

7.13

Why is vibration a serious health issue?

A) There are no early warning signs

B) The long-term effects of vibration are not known

C) There is no way that exposure to vibration can be prevented

D) Vibration can cause a disabling injury that cannot be cured

7.14

What are THREE early signs of vibration white finger or hand-arm vibration syndrome (HAVS)?

A) Temporary loss of feeling in the fingers

B) Fingertips turn white

C) Rash on fingers

D) Tingling or pins and needles sensation in the fingers

E) Blisters

7.15

Hand-arm vibration syndrome (HAVS) can cause:

A) skin cancer

B) skin irritation, like dermatitis

C) blisters on your hands and arms

D) damaged blood vessels and nerves in your fingers and hands

7.16

You have been using a vibrating tool. The end of your fingers are starting to tingle. What does this mean?

A) You can carry on using the tool but you must loosen your grip

B) You must not use this tool, or any other vibrating tool, ever again

C) You need to report your symptoms before they cause a problem

D) You can carry on using the tool but you must hold it tighter

7.17

What is vibration white finger or hand-arm vibration syndrome (HAVS)?

A) A mild skin rash that will go away

B) A serious skin condition that will not clear up

C) Severe frostbite

D) A sign that your hands and arms have or are on the way to being permanently damaged

7.18

Which of these is most likely to cause vibration white finger?

A) Handsaw

B) Hammer drill

C) Hammer and chisel

D) Battery-powered screwdriver

Answers: 7.13 = D   7.14 = A, B, D   7.15 = D   7.16 = C   7.17 = D   7.18 = B

7.19

You are likely to suffer LESS from hand-arm vibration if you are:

A very cold but dry

B cold and wet

C warm and dry

D very wet but warm

7.20

If you need to use a vibrating tool, even for a short time, how can you help reduce the risk of hand-arm vibration?

A Do not grip the tool too tightly

B Hold the tool away from you, at arm's length

C Use more force

D Hold the tool more tightly

7.21

If you have to use a vibrating tool, what would you expect your supervisor to do?

A Measure the level of vibration while you use the tool

B Explain the risk assessment and the safest way and length of time each day that you can use the tool

C Watch you use the tool to assess the level of vibration

D Help you to make up your own safe system of work

7.22

If you have to use a vibrating tool, how can you help reduce the effects of hand-arm vibration?

A Hold the tool tightly

B Do the work in short spells

C Do the job in one long burst

D Only use one hand on the tool at a time

B
07

B
08

8.1

What equipment should you have if you are doing non-licensed work on asbestos-containing materials?

- A Disposable overalls ('type 5')
- B Suitable respiratory protective equipment (RPE) (e.g. disposable face mask with a FFP3 rating)
- C Laceless footwear
- D All of these answers

8.2

Exposure to asbestos fibres may result in which disease?

- A Mesothelioma
- B Asbestosis
- C Lung cancer
- D All of these answers

8.3

After asbestos, which of the following causes the most ill health to construction workers?

- A Wood and MDF dust
- B Diesel fumes
- C Silica dust
- D Resin, solvent and paint vapours

8.4

Where might you come across asbestos?

- A In a house built between 1950 and 1990
- B In any building built or refurbished before the year 2000
- C In industrial buildings built between 1920 and 1990
- D Asbestos has now been removed from all buildings

8.5

How can asbestos be correctly identified?

- A The dust gives off a strong smell
- B By getting a sample analysed in a lab
- C By the colour of the dust
- D By putting a piece in water and seeing if it dissolves

8.6

What training do you need to work with or remove asbestos cement products?

- A General asbestos awareness training
- B Having a CSCS card tells me all I need to know
- C Training for non-licensable asbestos work
- D None – anyone can work with asbestos cement

8.7

If you think you have found some asbestos, the first thing you should do is:

A stop work and warn others

B take a sample to your supervisor

C put the bits in a bin and carry on with your work

D find the first aider

8.8

If you breathe in asbestos dust it can cause:

A aching muscles and painful joints

B throat infections

C lung diseases

D dizziness and headaches

8.9

This sign should be used for labelling:

A asbestos waste

B raw asbestos

C any product containing asbestos

D all of these answers

8.10

You find an unmarked container that you think might contain chemicals. What is the first thing you should do?

A Smell the chemical to see what it is

B Put it in a bin to get rid of it

C Ensure it remains undisturbed and report it

D Taste the chemical to see what it is

B
08

8.11

Cement bags have an additive to help prevent allergic dermatitis. When using a new bag what should be checked?

A The bag is undamaged

B The 'use by' date has not expired

C It has been stored in a dry place

D The contents are not hard and gone off

8.12

Why should you not kneel in wet cement, screed or concrete?

A It will make your trousers wet

B It is not an effective way to work

C It can cause serious chemical burns to your legs

D It will affect the finish

B
08

8.13

Wet cement, mortar and concrete is hazardous to your health as it causes:

A dizziness and headaches

B chemical burns and dermatitis

C muscle aches

D arc eye

8.14

Which of these will give you health and safety information about a hazardous substance?

A The site diary

B The delivery note

C The COSHH assessment ✓

D The accident book

8.15

You need to use a hazardous substance. Who should explain the health risks and safe method of work you need to follow (the COSHH assessment) before you start?

A A Health and Safety Executive (HSE) inspector

B The site first aider

C Your supervisor or employer ✓

D The site security people

8.16

A COSHH assessment tells you how:

A to lift heavy loads and how to protect yourself

B to work safely in confined spaces

C a substance might harm you and how to protect yourself when you are using it ✓

D noise levels are assessed and how to protect your hearing

8.17

The safest way to use a hazardous substance is to:

A get on with the job as quickly as possible

B read your employer's health and safety policy

C understand the COSHH assessment and follow the instructions ✓

D ask someone who has already used it

8.18

 If you see either of these labels on a substance what should you do?

A Do not use it as the substance is poisonous

B Find out what protection you need as the substance is corrosive and can damage your skin upon contact ✓

C Wash your hands after you have used the substance

D Find out what hand cleaner you will need as the substance will not wash off easily

8.19

If you see either of these labels on a substance what should you do?

(A) Find out what protection you need as the substance is harmful and could damage your health

(B) Use sparingly as substance is expensive

(C) Wear gloves as the substance can burn your skin

(D) Do not use it as the substance is poisonous

8.20

How can you tell if a product is hazardous?

(A) By warning symbols on the container or packaging label

(B) By the shape of the container

(C) It will always be in a black container

(D) It will always be in a cardboard box

8.21

The packaging of a substance has the word 'SENSITISER' on it. This means that:

(A) you could become allergic to it and have allergic reactions

(B) it must be mixed with water before you can use it

(C) it is perfectly safe to use without personal protective equipment (PPE)

(D) it should not be used under any circumstances

8.22

If you see either of these labels on a substance what should you do?

(A) Make sure it is stored out of the reach of children

(B) Use the substance very carefully and make sure you don't spill or splash it on yourself

(C) Do not use it as the substance is poisonous

(D) Find out what protection you need as the substance is toxic and in low quantities could seriously damage your health or kill you

8.23

If you see either of these labels on a substance what should you do?

(A) Find out how to handle the substance as it is fragile

(B) Find out how to use the substance safely as it could explode

(C) Find out how to use the substance safely as it could catch fire easily

(D) Do not use the substance as it could kill you

B 08

Answers: 8.19 = A 8.20 = A 8.21 = A 8.22 = D 8.23 = B

8.24

If you see either of these labels on a substance what should you do?

A Dispose of the substance or contents by burning

B Find out how to use the substance safely as it could explode

C Find out how to use the substance safely as it is flammable (could catch fire easily)

D Warm up the contents first, with heat or a naked flame

8.25

What does this warning sign mean?

A Substance can explode

B Substance will cause heartburn if swallowed

C Substance can glow in the dark

D Substance can cause long-term serious health problems

9.1

To lift a load safely you need to think about:

A) its size and shape

B) its weight

C) how to grip or hold it firmly

D) all of these answers

9.2

You are using a trolley to move a heavy load a long distance and a wheel comes off. What should you do?

A) Carry the load the rest of the way

B) Ask someone to help you pull the trolley the rest of the way

C) Drag the trolley on your own the rest of the way

D) Find another way to move the load

9.3

You have to move a load that might be too heavy for you. You cannot divide it into smaller parts and there is no-one to help you. What should you do?

A) Do not move the load until you have found a safe method

B) Get a forklift truck, even though you can't drive one

C) Try to lift it using the correct lifting methods

D) Lift and move the load quickly to avoid injury

9.4

You need to move a load that might be too heavy for you. What should you do?

A) Divide the load into smaller loads if possible

B) Get someone to help you

C) Use an aid, such as a trolley or wheelbarrow

D) All of these answers

9.5

You have to lift a heavy load. What must your employer do?

A) Make sure your supervisor is there to advise while you lift

B) Do a risk assessment of the task

C) Nothing, it is part of your job to lift loads

D) Watch you while you lift the load

9.6

You need to lift a load that is not heavy, but it is so big that you cannot see in front of you. What should you do?

A) Ask someone to help carry the load so that you can both see ahead

B) Get someone to walk next to you and give directions

C) Get someone to walk in front of you and tell others to get out of the way

D) Move the load on your own because it is so large that anyone in your way is sure to see it

B
09

Answers: 9.1 = D   9.2 = D   9.3 = A   9.4 = D   9.5 = B   9.6 = A

B
09

9.7

Who should be involved in creating the safe system of work for your manual handling?

- A You
- B Your supervisor/employer
- C You and your supervisor/employer ✓
- D The Health and Safety Executive (HSE)

9.8

You have to carry a load down a steep slope. What should you do?

- A Walk backwards down the slope to improve your balance
- B Carry the load on your shoulder
- C Assess whether you can still carry the load safely ✓
- D Run down the slope to finish quickly

9.9

Under the regulations for manual handling, all workers must:

- A wear back-support belts when lifting anything
- B make a list of all the heavy things they have to carry
- C lift any size of load once the risk assessment has been done
- D follow their employer's safe systems of work ✓

9.10

You are using a wheelbarrow to move a heavy load. Is this manual handling?

- A No, because the wheelbarrow is carrying the load
- B Only if the load slips off the wheelbarrow
- C Yes, you are still manually handling the load ✓
- D Only if the wheelbarrow has a flat tyre

9.11

Which part of your body is most likely to be injured if you lift heavy loads?

- A Your knees
- B Your back ✓
- C Your shoulders
- D Your elbows

9.12

You have been told how to lift a heavy load, but you think there is a better way to do it. What should you do?

- A Ignore what you have been told and do it your way
- B Ask your workmates to decide which way you should do it
- C Discuss your idea with your supervisor ✓
- D Forget your idea and do it the way you have been told

9.13

Your new job involves some manual handling. An old injury means that you have a weak back. What should you do?

A Tell your supervisor you can lift anything

B Tell your supervisor that lifting might be a problem

C Try some lifting then tell your supervisor about your back

D Tell your supervisor about your back if it gets injured again

9.14

Before you lift a heavy load you should always try to:

A stand with your feet together when lifting

B bend your back when lifting

C carry the load away from your body, at arm's length

D divide large loads into smaller loads or use lifting equipment

9.15

You need to lift a load from the floor. You should stand with your:

A feet together, legs straight, back bent

B feet together, knees bent, in a deep squatting position

C feet slightly apart, one leg slightly forward, knees bent

D feet wide apart, legs straight, back bent

9.16

If you have to twist or turn your body when you lift and place a load, it means:

A the weight you can lift safely is LESS than usual

B the weight you can lift safely is MORE than usual

C nothing, you can lift the SAME weight as usual

D you MUST wear a back brace

B
09

9.17

If you wear a back support belt when lifting:

A you can lift any load without being injured

B you can safely lift more than usual

C you could face the same risk of injury as when you are not wearing one

D it will crush your backbone and damage it

9.18

You need to move a load that is heavier on one side than the other. How should you pick it up?

A With the heavy side towards you

B With the heavy side away from you

C With the heavy side on your strong arm

D With the heavy side on your weak arm

Answers: 9.13 = B   9.14 = D   9.15 = C   9.16 = A   9.17 = C   9.18 = A

9.19

You have to move a load while you are sitting, not standing. How much can you lift safely?

A Less than usual

B The usual amount

C Twice the usual amount

D Three times the usual amount

9.20

You need to reach above your head and lower a load to the floor. Which of these is NOT true?

A It will be more difficult to keep your back straight and chin tucked in

B You will put extra stress on your arms and your back

C You can safely handle more weight than usual

D The load will be more difficult to control

9.21

Adopting safe manual handling methods will help you to:

A protect your back and reduce the risk of injury

B increase your strength

C leave work earlier that day

D lift heavier loads

9.22

The MAIN reason for ensuring safe manual handling techniques in the workplace is to:

A complete the job quickly

B prevent personal injury

C keep fit

D create additional work

9.23

Which of the following is a manual handling task?

A Lifting materials with a crane

B Climbing ladders or stairs

C Activities involving pushing, pulling, lowering and lifting

D Getting in and out of a vehicle used at work

9.24

Which of the following is the BEST method to help minimise the risk of injury when manual handling?

A Safely using lifting aids

B Making the site flatter before performing the task

C Asking someone else to carry the load

D Employing stronger people

9.25

What is the MAIN reason for using lifting aids when undertaking a manual handling activity?

A They help reduce the risk of personal injury

B You do not require training to use them

C You can lift any load with a lifting aid

D Lifting aids are expensive and should be used

9.26

What TWO things are important for the use of manual handling lifting aids?

A The user must hold a CSCS card

B The lifting aid can only be used outside

C The lifting aid must be designed for the task

D The lifting aid must not be more than six months old

E The user must be trained in the correct use of the lifting aid

B
09

Answers: 9.25 = A   9.26 = C, E

Manual handling

C

General safety

Contents

10.1

A crane has to do a difficult lift. The signaller asks you to help, but you are not trained in plant signals. What should you do?

- [A] Politely refuse and explain you don't know how to signal
- [B] Start giving signals to the crane driver
- [C] Only help if the signaller really can't manage alone
- [D] Ask the signaller to show you what signals to use

10.2

A truck has to tip materials into a trench. Who should give signals to the truck driver?

- [A] Anyone who is wearing a hi-vis coat
- [B] Someone standing in the trench
- [C] Someone who knows the signals
- [D] Only the person who is trained and appointed for the job

10.3

 What does this sign mean?

- [A] Assemble here in the event of a fire
- [B] Fire extinguishers and fire-fighting equipment kept here
- [C] Parking reserved for emergency service vehicles
- [D] Do not store flammable materials here

10.4

 What does this sign mean?

- [A] Fire alarm call point
- [B] Hot surface, do not touch
- [C] Wear flameproof hand protection
- [D] Emergency light switch

10.5

 What does this sign mean?

- [A] Press here to sound the fire alarm
- [B] Fire hose reel located here
- [C] Turn key to open fire door
- [D] Do not use if there is a fire

10.6

What does this sign mean?

- [A] Wear hearing protection if you want to
- [B] You must wear hearing protection
- [C] No personal stereos or MP3 players
- [D] Caution, noisy machinery

C
10

10.7

 What does this sign mean?

A Safety glasses cleaning station

B Warning, bright lights or lasers

C Caution, poor lighting

D You must wear safety eye protection

10.8

 Blue and white signs are:

A mandatory signs – meaning you MUST do something

B prohibition signs – meaning you MUST NOT do something

C warning signs – alerting you to hazards or danger

D safe condition signs – giving you information

10.9

 What does this sign mean?

A Safety boots or safety shoes must be worn

B Wellington boots must be worn

C Be aware of slip and trip hazards

D No dirty footwear past this point

10.10

 What does this sign mean?

A You must carry safety gloves at all times

B Dispose of used safety gloves here

C Safety gloves do not need to be worn

D Safety gloves must be worn

10.11

 What does this sign mean?

A Wear white clothes at night

B Hi-vis clothing must be worn

C Do nothing, it only applies to managers

D Cover up bare arms

C 10

10.12

 What does this sign mean?

A Smoking is allowed

B Danger flammable materials present

C No smoking

D No explosives or naked flames

Answers: 10.7 = D   10.8 = A   10.9 = A   10.10 = D   10.11 = B   10.12 = C

10.13

 What does this sign mean?

- A. No lone working
- B. No entry without a hard hat
- C. No pedestrians or entry for people on foot
- D. No entry during the day

10.14

Round red and white signs with a diagonal line are:

- A. mandatory signs – meaning you MUST do something
- B. prohibition signs – meaning you MUST NOT do something
- C. warning signs – alerting you to hazards or danger
- D. safe condition signs – giving you information

10.15

What does this sign mean?

- A. Do not jump across any gaps in the scaffold
- B. Do not work on the first lift of the scaffold
- C. Do not access the scaffold because it is incomplete or not safe
- D. Do not walk under the scaffold

10.16

 What does this sign mean?

- A. No running allowed
- B. There is no escape route
- C. This is a fire door
- D. Fire escape route

10.17

 What does this sign mean?

- A. It tells you where the canteen is located
- B. It tells you which direction to walk
- C. It tells you where to assemble in case of an emergency
- D. It tells you where the site induction room is located

10.18

Green and white signs are:

- A. mandatory signs – meaning you MUST do something
- B. prohibition signs – meaning you MUST NOT do something
- C. warning signs – alerting you to hazards or danger
- D. safe condition signs – giving you information

Answers: 10.13 = C   10.14 = B   10.15 = C   10.16 = B   10.17 = C   10.18 = D

10.19

 What does this sign mean?

(A) Toilets and shower facilities

(B) Drying area for wet weather clothes

(C) Emergency first-aid shower

(D) Fire sprinklers above

10.20

 What does this sign mean?

(A) Risk of electrocution

(B) Risk of static shock

(C) Live electrical appliance

(D) Risk of lightning

10.21

 What does this sign mean?

(A) Dispose of substance or contents by burning

(B) Warning – substance or contents are flammable (can catch fire easily)

(C) Warning – substance or contents could explode

(D) Warning – substance or contents are harmful

10.22

 What does this sign mean?

(A) Radioactive area

(B) Warning – explosive substance

(C) Flashing lights ahead

(D) Warning – laser beams

10.23

 Yellow and black signs are:

(A) mandatory signs – meaning you MUST do something

(B) prohibition signs – meaning you MUST NOT do something

(C) warning signs – alerting you to hazards or danger

(D) safe condition signs – giving you information

10.24

 What does this sign mean?

(A) Plant operators are wanted

(B) Industrial vehicles are moving about

(C) Manual handling is not allowed

(D) Storage area

C
10

11.1

A fire assembly point is the place where:

- A. fire engines must go when they arrive on site
- B. the fire extinguishers are kept
- C. people must go when the fire alarm sounds
- D. the fire started

11.2

If you discover a fire, the first thing you should do is:

- A. put your tools away
- B. finish what you are doing, if it is safe to do so
- C. try to put out the fire
- D. raise the alarm

11.3

If you hear the fire alarm, you should go to the:

- A. site canteen
- B. assembly point
- C. site office
- D. fire

11.4

A large fire has been reported. You have not been trained to use fire extinguishers. You should:

- A. put away all your tools and then go to the assembly point
- B. report to the site office and then go home
- C. go straight to the assembly point
- D. leave work for the day

11.5

What does a hot work permit NOT tell you?

- A. When you can start and when you must stop the hot work
- B. How you must prevent sparks or heat travelling
- C. Where the local fire station is located
- D. What fire extinguisher or fire watch you need

11.6

A hot work permit allows you to:

- A. work in hot weather
- B. carry out work that needs warm, protective clothing
- C. carry out work that could start a fire
- D. light a bonfire

C
11

Answers: 11.1 = C   11.2 = D   11.3 = B   11.4 = C   11.5 = C   11.6 = C

11.7

If your job needs a hot work permit, what TWO things would you expect to have to do?

A Have a fire extinguisher close to the work

B Check for signs of fire when you stop work

C Know where all the fire extinguishers are kept on site

D Write a site evacuation plan

E Know how to refill fire extinguishers

11.8

Look at these jobs. Which TWO are likely to need a hot work permit?

A Cutting steel with an angle grinder

B Soldering pipework in a central heating system

C Refuelling a diesel dump truck

D Replacing an empty liquefied petroleum gas cylinder with a full one

E Using the heaters in the drying room

11.9

A fire needs heat, fuel and:

A oxygen

B carbon dioxide

C argon

D nitrogen

11.10

When you use a carbon dioxide (CO_2) fire extinguisher, the nozzle gets:

A very cold

B very hot

C warm

D very heavy

11.11

Which TWO extinguishers should NOT be used on electrical fires?

A Dry powder (Blue colour band)

B Foam (Cream colour band)

C Water (Red colour band)

D Carbon dioxide (Black colour band)

11.12

A WATER fire extinguisher, identified by a red band, should ONLY be used on what type of fire?

A Wood, paper, textile and solid material fires

B Flammable liquids (fuel, oil, varnish, paints, etc.)

C Electrical fires

D Metal and molten metal

C 11

Answers: 11.7 = A, B   11.8 = A, B   11.9 = A   11.10 = A   11.11 = B, C   11.12 = A

11.13

A DRY POWDER fire extinguisher, identified by a blue band, could be used on all types of fire but is BEST suited to what TWO types of fire?

A ☐ Wood, paper, textile and solid material fires

B ☐ Flammable liquids (fuel, oil, varnish, paints, etc.)

C ☐ Flammable gas (LPG, propane, etc.)

D ☐ Metal and molten metal

E ☐ Electrical fires

11.14

A FOAM extinguisher, identified by a cream band, should NOT be used on what type of fire?

A ☐ Wood, paper, textile and solid material fires

B ☐ Flammable liquids (fuel, oil, varnish, paints, etc.)

C ☐ Metal and molten metal

D ☐ Fires with cooking appliances

11.15

A CARBON DIOXIDE (CO_2) extinguisher, identified by a black band, should NOT be used on what type of fire?

A ☐ Wood, paper, textile and solid material fires

B ☐ Flammable liquids (fuel, oil, varnish, paints, etc.)

C ☐ Electrical fires

D ☐ Metal and molten metal

11.16

If you see 'frost' around the valve on a liquefied petroleum gas (LPG) cylinder, it means:

A ☐ the cylinder is nearly empty

B ☐ the cylinder is full

C ☐ the valve is leaking

D ☐ you must lay the cylinder on its side

11.17

If there is a fire you will need to go to the site assembly point. How would you expect to find out where this is?

A ☐ During a visit by the Health and Safety Executive (HSE)

B ☐ During site induction

C ☐ By reading your employer's health and safety policy

D ☐ Your colleagues will tell you

C
11

11.18

What is the MAIN aim of fire precautions on site?

A To have measures in place to fight a fire

B For the emergency services to reach a fire as quickly as possible

C To ensure everyone reaches safety in the event of a fire

D To prevent a fire spreading

11.19

How can you help PREVENT a fire hazard?

A Store solvents and paints in the drying room

B Leave your clothes over a heater all night

C Keep your work area tidy and place waste in the bins provided

D Store materials and equipment along the exit routes

11.20

You need to work in a corridor that is a fire escape route. You must see that:

A your tools and equipment do not block the route

B all doors into the corridor are locked

C you only use spark-proof tools

D you remove all fire escape signs before you start

11.21

What are TWO common fire risks on construction sites?

A 230 volt power tools

B Poor housekeeping and build up of waste

C Timber racks

D Uncontrolled hot works

E 110 volt extension reels

C
11

Answers: 11.18 = C   11.19 = C   11.20 = A   11.21 = B, D

12.1

You need to use an extension cable. What TWO things must you do?

A. Only uncoil the length of cable you need

B. Uncoil the whole cable

C. Clean the whole cable with a damp cloth

D. Check the whole cable and connectors for damage

E. Only check the cable you need for damage

12.2

You need to run an electrical cable across an area used by vehicles. What TWO things should you do?

A. Wrap the cable in yellow tape so that drivers can see it

B. Cover the cable with a protective ramp

C. Cover the cable with scaffold boards

D. Put up a sign that says 'Ramp ahead'

E. Run the cable at head height

12.3

You need to work near an electrical cable. The cable has bare wires. What should you do?

A. Quickly touch the cable to see if it is live

B. Check there are no sparks coming from the cable and then start work

C. Tell your supervisor and keep well away

D. Push the cable out of the way so that you can start work

12.4

If an extension cable has a cut in its outer cover, you should:

A. check the copper wires aren't showing in the cut and then use the cable

B. put electrical tape around the damaged part

C. report the fault and make sure no-one else uses the cable

D. put a bigger fuse in the cable plug

12.5

What is the best way to protect an extension cable while you work, as well as minimising trip hazards?

A. Run the cable above head height

B. Run the cable by the shortest route

C. Cover the cable with yellow tape

D. Cover the cable with pieces of wood

C
12

Answers: 12.1 = B, D   12.2 = B, D   12.3 = C   12.4 = C   12.5 = A

12.6

Untidy leads and extension cables are responsible for many trips and lost work time injuries. What TWO things should you do to help?

A Run cables and leads above head height and over the top of doorways and walkways rather than across the floor

B Tie any excess cables and leads up into the smallest coil possible

C Keep cables and leads close to the wall and not in the middle of the floor or walkway

D Make sure your cables go where you want them to and not worry about others

E Unplug the nearest safety lighting and use these sockets instead

12.7

To operate a powered hand tool, you must be:

A over 16 years old

B over 18 years old

C trained and competent

D 21 years old or over

12.8

You must be fully trained before you use a cartridge-operated tool. Why?

A They are heavy and could cause manual handling injuries

B They operate like a gun and can be dangerous in inexperienced hands

C Using one can cause dermatitis

D They have exposed electrical parts

12.9

If the guard is missing from a power tool you should:

A try to make another guard

B use the tool but try to work quickly

C not use the tool until a proper guard has been fitted

D use the tool but work carefully and slowly

12.10

If you need to use a power tool with a rotating blade, you should:

A remove the guard so that you can clearly see the blade

B adjust the guard to expose just enough blade to let you do the job

C remove the guard but wear leather gloves to protect your hands

D adjust the guard to expose the maximum amount of blade

12.11

Most cutting and grinding machines have guards. What are the TWO main functions of the guard?

A To stop materials getting onto the blade or wheel

B To give you a firm handhold

C To balance the machine

D To stop fragments flying into the air

E To stop you coming into contact with the blade or wheel

C
12

12.12

If you need to use a hand tool or power tool on site it must be:

[A] made in the UK

[B] the right tool for the job and inspected at the start of each week

[C] bought from a builders merchant

[D] the right tool for the job and inspected before you use it

12.13

Before you adjust an electric hand tool you should:

[A] switch it off but leave the plug in the socket

[B] switch it off and remove the plug from the socket

[C] do nothing in particular

[D] put tape over the ON/OFF switch

12.14

If the head on your hammer comes loose you should:

[A] stop work and get the hammer repaired or replaced

[B] find another heavy tool to use instead of the hammer

[C] keep using it but be aware that the head could come off at any time

[D] tell the other people near you to keep out of the way

12.15

Do you need to inspect simple hand tools like trowels, screwdrivers, saws and hammers?

[A] No, never

[B] Yes, if they have not been used for a couple of weeks

[C] Yes, they should be checked each time you use them

[D] Only if someone else has borrowed them

12.16

Someone near you is using a rotating laser level. What, if any, is the health hazard likely to affect you?

[A] Skin cancer

[B] None – if used correctly they are safe

[C] Gradual blindness

[D] Burning of the skin, similar to sunburn

12.17

What is the main danger if you use a chisel or bolster with a 'mushroomed' head?

[A] It will shatter and send fragments flying into the air

[B] It will damage the face of the hammer

[C] The shaft of the chisel will bend when you hit it

[D] You will have to sharpen the chisel more often

C
12

12.18

Look at these statements about power tools. Which one is true?

A Always carry the tool by its cord

B Always unplug the tool by pulling its cord

C Always unplug the tool when you are not using it ,

D Always leave the tool plugged in when you check or adjust it

12.19

It is dangerous to run an abrasive wheel faster than its recommended top speed. Why?

A The wheel will get clogged and stop

B The motor could burst into flames

C The wheel could shatter and burst into many pieces ,

D The safety guard cannot be used

12.20

It is safe to work close to an overhead power line if:

A you do not touch the line for more than 30 seconds

B you use a wooden ladder

C the power is switched off ,

D it is not raining

12.21

You are using electric equipment when it cuts out. You should:

A shake it to see if it will start again

B pull the electric cable to see if it is loose

C switch the power off and on a few times

D switch off the power and look for signs of damage ,

12.22

Someone near you is using a disc cutter to cut concrete blocks. What THREE immediate hazards are likely to affect you?

A Flying fragments ,

B Dermatitis

C Harmful dust in the air ,

D High noise levels ,

E Vibration white finger

C
12

12.23

You need to use an air-powered tool. Which of these is NOT a hazard?

A Electric shock ,

B Hand-arm vibration

C Airborne dust and flying fragments

D Leaking hoses

C
12

12.24

When do you need to check tools and equipment for damage?

- A Each time before use
- B Every day
- C Once a week
- D At least once a year

12.25

What are the TWO main areas of visual inspections you should carry out before each use of a power tool?

- A Check the carry case isn't broken
- B Check the power lead, plug and casing are in good condition
- C Check the manufacturer's label hasn't come off
- D Check switches, triggers and guards are adjusted and work correctly
- E Check if there is an upgraded model available

12.26

You should use a RCD (residual current device) with 230 volt tools because it:

- A lowers the voltage
- B quickly cuts off the power if there is a fault
- C makes the tool run at a safe speed
- D saves energy and lowers costs

12.27

How do you check if a RCD (residual current device) connected to a power tool is working?

- A Switch the tool on and off
- B Press the test button on the RCD (residual current device)
- C Switch the power on and off
- D Run the tool at top speed to see if it cuts out

12.28

You need to use a 230 volt item of equipment. How should you protect yourself from an electric shock?

- A Use a generator
- B Put up safety screens around you
- C Use a portable RCD (residual current device)
- D Wear rubber boots and gloves

12.29

The Portable Appliance Testing (PAT) label on a power tool tells you:

- A when the next safety check is due
- B when the tool was made
- C who tested the tool before it left the factory
- D its earth-loop impedance

12.30

On building sites, the recommended safe voltage for electrical equipment is:

A) 12 volts

B) 24 volts

C) 110 volts

D) 230 volts

12.31

The colour of a 110 volt power cable and connector should be:

A) black

B) red

C) blue

D) yellow

12.32

Why should you try to use battery-powered tools rather than electrical ones?

A) They are cheaper to run

B) They will not give you an electric shock

C) They will not give you hand-arm vibration

D) They do not need to be tested or serviced

12.33

Why do building sites use a 110 volt electricity supply instead of the usual 230 volt domestic supply?

A) It is cheaper

B) It is less likely to kill you

C) It moves faster along the cables

D) It is safer for the environment

C
12

Answers: 12.30 = C   12.31 = D   12.32 = B   12.33 = B

13.1

What are the TWO conditions for being able to operate plant on site?

A You must be trained and competent

B You must be authorised

C You must be over 21 years old

D You must hold a full driving licence

E You must hold a UK passport

13.2

Your supervisor asks you to drive a dumper truck. You have never driven one before. What should you do?

A Ask a trained driver how to operate it

B Explain to your supervisor that you are not trained and therefore cannot operate it

C Watch other dumpers to see how they are operated

D Get on with it

13.3

A lorry is in trouble as it tries to reverse into a tight space. You have not been trained as a signaller. What should you do?

A Stay well out of the way

B Help the driver by giving hand signals

C Help the driver by jumping up into the cab

D Offer to adjust the mirrors on the lorry

13.4

What do you need before you can supervise any lift using a crane?

A Nothing, you make it up as you go along

B You must be trained and assessed as competent

C Written instructions from the crane hire company

D Nothing, the crane driver will tell you what to do

13.5

A mobile plant operator can let people ride in or on the machine:

A if they have a long way to walk

B as long as the site speed limit is not exceeded

C only if it is designed to carry passengers and has a designated seat

D if the cab door is shut

13.6

You need to walk past someone using a mobile crane. You should:

A guess what the crane operator will do next and squeeze by

B try to catch the attention of the crane operator

C run to get past the crane quickly

D take another route so that you stay clear of the crane

13.7

When you walk across the site, what is the best way to avoid an accident with mobile plant?

A Keep to the designated pedestrian routes

B Ride on the plant

C Get the attention of the driver before you get too close

D Wear hi-vis clothing

13.8

You need to walk past a 360° mobile crane. The crane is operating near a wall. What is the main danger?

A The crane could crash into the wall

B You could be crushed if you walk between the crane and the wall

C Whole body vibration from the crane

D High noise levels from the crane

13.9

You are walking across the site. A large mobile crane reverses across your path. What should you do?

A Help the driver to reverse

B Start to run so that you can pass behind the reversing crane

C Pass close to the front of the crane

D Wait or find another way around the crane

13.10

When is site transport allowed to drive along a pedestrian route?

A During meal breaks

B If it is the shortest route

C Only if necessary and if all pedestrians are excluded

D Only if the vehicle has a flashing yellow light

13.11

Which of these would you NOT expect to see if site transport is well organised?

A Speed limits

B Barriers to keep pedestrians away from mobile plant and vehicles

C Pedestrians and mobile plant using the same routes

D One-way systems

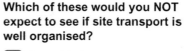

C
13

13.12

How would you expect a well-organised site to keep pedestrians away from traffic routes?

A The site manager will direct all pedestrians away from traffic routes

B The traffic routes will be shown on the Health and Safety Law poster

C There will be barriers between traffic and pedestrian routes

D There is no need to keep traffic and pedestrians apart

Answers: 13.7 = A   13.8 = B   13.9 = D   13.10 = C   13.11 = C   13.12 = C

C
13

13.13

A site vehicle is most likely to injure pedestrians when it is:

- A reversing
- B lifting materials onto scaffolds
- C tipping into an excavation
- D digging out footings

13.14

You must not walk behind a lorry when it is reversing because:

- A most lorries are not fitted with mirrors
- B the driver is unlikely to know you are there
- C most lorry drivers aren't very good at reversing
- D you will need to run, not walk, to get past it in time

13.15

The quickest way to your work area is through a contractor's vehicle compound. Which way should you go?

- A Around the compound if vehicles are moving
- B Straight through the compound if no vehicles appear to be moving
- C Around the compound every time
- D Straight through the compound if no-one is looking

13.16

A forklift truck is blocking the way to where you want to go on site. It is lifting materials onto a scaffold. What should you do?

- A Only walk under the raised load if you are wearing a safety helmet
- B Catch the driver's attention and then walk under the raised load
- C Start to run so that you are not under the load for very long
- D Wait or go around, but never walk under a raised load

13.17

You think some mobile plant is operating too close to where you have to work. What should you do first?

- A Stop work and speak to the plant operator
- B Stop work and speak to the plant operator's supervisor
- C Keep a good lookout for the plant and carry on working
- D Stop work and speak to your own supervisor

13.18

If you see a dumper being driven too fast you should:

- A keep out of its way and report the matter
- B try to catch the dumper and speak to the driver
- C report the matter to the police
- D do nothing, dumpers are allowed to go above the site speed limit

Answers: 13.13 = A   13.14 = B   13.15 = C   13.16 = D   13.17 = D   13.18 = A

13.19

You see a lorry parking. It has a flat tyre. Why should you tell the driver?

A The lorry will use more fuel

B The lorry will need to travel at a much slower speed

C The lorry could be unsafe to drive

D The lorry can only carry small loads

13.20

An excavator has just stopped work. Liquid is dripping and forming a small pool under the back of the machine. What could this mean?

A It is normal for fluids to vent after the machine stops

B The machine is hot so the diesel has expanded and overflowed

C Someone put too much diesel into the machine before it started work

D The machine has a leak and could be unsafe

13.21

You see a mobile crane lifting a load. The load is about to hit something. What should you do?

A Go and tell your supervisor

B Try and warn the person supervising or banking the lift

C Go and tell the crane driver

D Do nothing and assume everything is under control

13.22

You see a driver refuelling an excavator. Most of the diesel is spilling onto the ground. What is the first thing you should do?

A Tell your supervisor the next time you see them

B Tell the driver immediately

C Look for a spillage kit immediately

D Do nothing, the diesel will eventually seep into the ground

13.23

You think a load is about to fall from a moving forklift truck. What should you do?

A Keep clear but try to warn the driver and others in the area

B Run alongside the machine and try to hold on to the load

C Run and tell your supervisor

D Sound the nearest fire alarm bell

13.24

How would you expect to be told about the site traffic rules?

A During site induction

B By a Health and Safety Executive (HSE) inspector

C By a note on a noticeboard

D In a letter sent to your home

C
13

Site transport safety

D

High risk activities

Contents

14.1

Which type of accident kills most construction workers?

(A) Falling from height

(B) Contact with electricity

(C) Being run over by site transport

(D) Being hit by a falling object

14.2

If you store materials on a working platform, which statement is correct?

(A) Materials can be stored unsecured above the guard-rail height

(B) Materials must be stored so they can't fall and the platform must be able to take their weight

(C) Materials can be stored anywhere, even if they pose a trip hazard or block the walkway

(D) Materials do not need to be secured if they are going to be there for less than an hour

14.3

Working at height is:

(A) 1.2 m above the ground or higher

(B) 2 m above the ground or higher

(C) any height that would cause an injury if you fell

(D) 3 m above the ground or higher

14.4

Which of the following is NOT true when using podium steps?

(A) The 'wheels' must be locked before you get on to them

(B) Podiums can easily topple if you over reach sideways

(C) Podiums are safe and can't topple over

(D) Podiums are work equipment and must be inspected every seven days

14.5

A ladder should not be painted because:

(A) the paint will make it slippery to use

(B) the paint may hide any damaged parts

(C) the paint could damage the metal parts of the ladder

(D) it will need regular re-painting

14.6

How many people should be on a ladder at the same time?

(A) One

(B) Two

(C) One on each section of an extension ladder

(D) Three, if it is long enough

D
14

Answers: 14.1 = A   14.2 = B   14.3 = C   14.4 = C   14.5 = B   14.6 = A

14.7

You find a ladder that is damaged. What should you do?

A) Don't use it and make sure that others know about the damage

B) Don't use it and report the damage at the end of your shift

C) Try and mend the damage

D) Use the ladder if you can avoid the damaged part

14.8

When using a ladder what should the slope or angle of the ladder be?

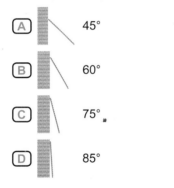

A) 45°

B) 60°

C) 75°

D) 85°

14.9

Who should check a ladder before it is used?

A) The person who is going to use it

B) A supervisor

C) The site safety officer

D) The manufacturer

14.10

What is the best way to make sure that a ladder is secure and won't slip?

A) Tie it at the top

B) Ask someone to stand with their foot on the bottom rung

C) Tie it at the bottom

D) Wedge the bottom of the ladder with blocks of wood

14.11

When could you use a ladder at work?

A) If it is long enough

B) If you can find a ladder to use

C) If other people do not need to use it for access

D) If you are doing light work for a short time

14.12

When you climb a ladder you must:

A) have three points of contact with the ladder at all times

B) have two points of contact with the ladder at all times

C) use a safety harness

D) have two people on the ladder at all times

D
14

14.13

You need to use a ladder to get to a scaffold platform. Which of these statements is true?

A. It must be tied and extend five rungs above the platform .

B. All broken rungs must be clearly marked

C. It must be wedged at the bottom to stop it slipping

D. Two people must be on the ladder at all times

14.14

You need to use a mobile tower scaffold. The wheel brakes do not work. What should you do?

A. Use some wood to wedge the wheels and stop them moving

B. Do not use the tower .

C. Only use the tower if the floor is level

D. Get someone to hold the tower while you use it

14.15

You need to reach the working platform of a mobile tower scaffold. What is the right way to do this?

A. Climb up the tower frame on the outside of the tower

B. Lean a ladder against the tower and climb up that

C. Climb up the ladder built into the tower ,

D. Climb up the outside of the diagonal bracing

14.16

A mobile tower scaffold must NOT be used on:

A. soft or uneven ground .

B. a paved patio

C. an asphalt road

D. a smooth concrete path

14.17

When working in a mobile elevating work platform (MEWP) where should you attach your harnesses lanyard?

A. To the control box

B. To a point on the structure or building you are working on

C. To the MEWP handrail

D. To the designated anchor point within the platform or basket ,

D 14

Answers: 14.13 = A   14.14 = B   14.15 = C   14.16 = A   14.17 = D

14.18

If you have to work at height over or near to deep water, which following item of personal protective equipment (PPE) must you be wearing?

A) Wellington boots

B) Lifejacket ,

C) Full face respirator

D) Full body harness

14.19

It is safe to cross a fragile roof if you:

A) walk along the line of bolts

B) can see fragile roof signs

C) don't walk on any plastic panels

D) use crawling boards with handrails,

14.20

You are working on a flat roof. What is the best way to stop yourself falling over the edge?

A) Put a large warning sign at the edge of the roof

B) Ask someone to watch you and shout when you get too close to the edge

C) Protect the edge with a guard-rail and toe-board ,

D) Use red and white tape to mark the edge

14.21

What is the best way to stop people falling through voids, holes or fragile roof panels?

A) Tell everyone where the dangerous areas are

B) Secure in place covers that can take the weight of a person and add warning signage ,

C) Cover them with netting

D) Mark the areas with red and white tape

14.22

 What does this sign mean?

A) Load-bearing roof. OK to stand on surface but not any roof lights

B) Fragile roof. Take care when walking on roof surface

C) Fragile roof. Do not stand directly on roof but use fall protection measures ,

D) Load-bearing roof. Surface can be slippery when wet

D
14

14.23

Who should erect, dismantle or alter a tube and fitting scaffold?

A) Anyone who thinks they can do it

B) Anyone who has the right tools

C) Anyone who is trained, competent and authorised ,

D) Anyone who is a project manager

14.24

You need to stack materials on a working platform. What is the best way to stop them falling over the toe-board?

A Fit brick guards or netting to the edge.

B Put a warning sign on the stack

C Build the stack so that it leans away from the edge

D Cover the stack with polythene

14.25

A scaffold guard-rail must be removed to allow you to carry out a survey. You are not a scaffolder. Can you remove the guard-rail?

A Yes, if you put it back as soon as you have finished

B Yes, if you put it back before you leave site

C No, only a scaffolder can remove the guard-rail but you can put it back

D No, only a scaffolder can remove the guard-rail and put it back.

14.26

Tools and materials can easily fall from a scaffold platform. What is the best way to protect the people below?

A Make sure they are wearing safety helmets

B Tell them you will be working above them

C Use brick guards to stop any items falling below.

D Tell the people below to stop work and clear the area

14.27

What must you NOT do when working at height?

A Follow the safe system of work

B Stop if you feel unsafe

C Take risks if you feel it is safe to do so.

D Use the safety equipment provided

14.28

You are required to use access equipment that you have not been trained to use. What should you do?

A Do the job as it will only take 10 minutes

B Get a ladder instead of using the correct access equipment

C Stop work and speak to your supervisor

D Ask someone else to do it

D
14

Answers: 14.24 = A   14.25 = D   14.26 = C   14.27 = C   14.28 = C

14.29

When working at height TWO of your responsibilities are to:

A) ensure you are sufficiently trained

B) make use of access equipment

C) throw things to your colleague below

D) ignore the safety briefing given by your supervisor

E) climb up the outside of the scaffolding

14.30

What is the MAIN regulation that controls the use of suitable equipment for working at height?

A) Control of Substances Hazardous to Health Regulations

B) Lifting Operations and Lifting Equipment Regulations

C) Work at Height Regulations

D) Workplace (Health, Safety and Welfare) Regulations

D
14

15.1

What is the safest way to get into a deep excavation?

A) Climb down a secured ladder

B) Use the buried services as steps

C) Climb down the shoring

D) Go down in an excavator bucket

15.2

You are in a deep trench. A lorry backs up to the trench and the engine is left running. What should you do?

A) Put on ear defenders to cut out the engine noise

B) Ignore the problem, the lorry will soon drive away

C) See if there is a toxic gas meter in the trench

D) Get out of the trench quickly

15.3

You are in a deep trench and start to feel dizzy. What should you do?

A) Get out, let your head clear and then go back in again

B) Carry on working and hope that the feeling will go away

C) Make sure that you and any others get out quickly and report it

D) Sit down in the trench and take a rest

15.4

Before work starts in a confined space, how should the air be checked?

A) Someone should go in and sniff the air

B) The air should be tested with a meter

C) Someone should look around to see if there is toxic gas

D) The air should be tested with a match to see if it stays alight

15.5

If there is sludge at the bottom of a confined space, you should:

A) go in and then step into the sludge to see how deep it is

B) throw something into the sludge to see how deep it is

C) put on a disposable facemask before you go in

D) have the correct respiratory protective equipment (RPE) and training before you go in

15.6

Why is methane gas dangerous in confined spaces? Give TWO answers.

A) It can explode

B) It makes you hyperactive

C) You will not be able to see because of the dense fumes

D) It makes you dehydrated

E) You may not have enough oxygen to breathe

D 15

15.7

You are in a confined space. If the level of oxygen drops:

A your hearing could be affected

B there is a high risk of fire or explosion

C you could become unconscious

D you might get dehydrated

15.8

You are working in a confined space when you notice the smell of bad eggs. This smell is a sign of:

A hydrogen sulphide

B oxygen

C methane

D carbon dioxide

15.9

You need to walk through sludge at the bottom of a confined space. Which of these is NOT a hazard?

A The release of oxygen

B The release of toxic gases

C Slips and trips

D The release of flammable gases

15.10

What is the main reason for having a person positioned immediately outside a confined space whilst work is taking place inside it?

A To supervise the work taking place inside the confined space

B To check compliance with the method statement

C To get the rescue plan underway in an emergency

D To carry out a risk assessment for the work

15.11

You are in a confined space when the gas alarm sounds. You have no respiratory protective equipment (RPE). What should you do?

A Switch off the alarm

B Get out of the confined space quickly

C Carry on working but do not use electrical tools

D Carry on working but take plenty of breaks in the fresh air

D 15

15.12

An excavation must be supported if:

A it is more than 5 m deep

B it is more than 1.2 m deep

C there is a risk of the sides falling in

D any buried services cross the excavation

15.13

You are working in an excavation. If you see the side supports move, you should:

- A) keep watching to see if they move again
- B) make sure that you and others get out quickly
- C) do nothing as the sides move all the time
- D) work in another part of the excavation

15.14

Guard-rails are placed around the top of an excavation to prevent:

- A) toxic gases from collecting in the bottom of the trench
- B) anyone falling into the trench and being injured
- C) the sides of the trench from collapsing
- D) rain water running off the ground at the top and into the trench

15.15

What must happen each time a shift starts work in an excavation?

- A) Someone must go in and sniff the air to see if it is safe
- B) A competent person must inspect the excavation
- C) A supervisor should stay in the excavation for the first hour
- D) A supervisor should watch from the top for the first hour

15.16

Work in a confined space usually needs three safety documents – a risk assessment, a method statement and:

- A) a permit to work/enter
- B) an up to date staff handbook
- C) a written contract for the work
- D) a company health and safety policy

15.17

Why is it important that people are trained before they are allowed to go into a confined space?

- A) Confined spaces never contain breathable air
- B) The conditions inside a confined space may be harmful to health
- C) Confined spaces are only found on house-building sites
- D) Confined spaces always contain flammable or explosive gases

15.18

You have to work in a confined space. There is no rescue team or rescue plan. What should you do?

- A) Assume that a rescue team or plan is not necessary and do the job
- B) Get someone to stand at the opening with a rope
- C) Do not enter until a rescue plan and team are in place
- D) Carry out the job in short spells

Answers: 15.13 = B   15.14 = B   15.15 = B   15.16 = A   15.17 = B   15.18 = C

15.19

You are working in a confined space. If the permit to work runs out before you can finish the job, you should:

A carry on working until the job is finished

B hand the permit over to the next shift

C ask your supervisor to change the date on the permit

D leave the confined space before the permit runs out

15.20

When digging, you notice the soil gives off a strange smell. What is this likely to mean?

A The soil contains a lot of clay

B The soil has been excavated before

C The ground has been used to grow crops in the past

D The ground could be contaminated

15.21

The main cause of death when people have to work in a confined space is:

A the presence of methane gas

B inadequate emergency rescue plan and equipment in place

C the disturbance of sludge

D too much oxygen

15.22

When digging, you hit and damage a buried cable. What should you do?

A Move the cable out of the way and carry on digging

B Wait 10 seconds and then move the cable out of the way

C Do not touch the cable, stop work and report it

D Dig round the cable or dig somewhere else

15.23

When digging you find a run of yellow plastic marker tape. What does it mean?

A There are buried human remains and you must tell your supervisor

B There is a buried service and further excavation must be carried out with care

C The soil is contaminated and you must wear respiratory protective equipment (RPE)

D The excavation now needs side supports

D
15

15.24

Which of these is the most accurate way to locate buried services?

A Cable plans

B Trial holes

C Survey drawings

D Architect drawings

15.25

If you need to dig near underground services, you should only be using:

A a jack hammer

B an insulated spade or shovel ِ

C a pick and fork

D an excavator

D
15

E

Environment

Contents

16.1

If you find bats on site, which of the following statements is true?

- (A) Bats are NOT a protected species so you can disturb or destroy their shelters or resting places
- (B) You can move the bats as long as you do it at night when they are out foraging
- (C) You can disturb or destroy shelters or resting places of bats if they get in the way of building work
- (D) Bats are a protected species so you cannot disturb or destroy their shelters or resting places *

16.2

You discover a bird on a nest where you need to work. What should you do?

- (A) Cover it with a bucket
- (B) Move it, do your work and then put it back
- (C) Make others aware of its presence whilst you go and inform your supervisor .
- (D) Scare it away

16.3

You are on site and you need to throw away some waste liquid that has oil in it. What should you do?

- (A) Pour it down a drain outside the site
- (B) Pour it onto the ground and let it soak away
- (C) Use it to start a fire
- (D) Find out how you should get rid of it from your supervisor or environmental advisor .

16.4

How should you get rid of hazardous waste?

- (A) If the waste has a hazardous symbol on it, then it can be put in any skip on site
- (B) Place it in the correctly labelled container or ask your supervisor .
- (C) Put it only in a mixed waste skip
- (D) Take it to the nearest local authority waste tip

16.5

Which of the following should be classed as hazardous waste?

- (A) Broken ceramic tiles or bricks
- (B) Polythene and shrink wrap
- (C) Glass
- (D) Fluorescent light tubes .

16.6

Which of the following should be disposed of as hazardous waste?

- (A) Softwood timber off-cuts
- (B) Glass fibre insulation
- (C) Part full tins of oil-based paint .
- (D) Damaged hard hat

E
16

Answers: 16.1 = D   16.2 = C   16.3 = D   16.4 = B   16.5 = D   16.6 = C

16.7

You have been asked to clean up oil that has leaked from machinery onto the ground. What is the right way to do this?

- A Put the oily soil into the general waste skip
- B Put the oily soil into a separate container for collection as hazardous waste
- C Mix the soil up with other soil so that the oil cannot be seen
- D Wash the oil away with water and detergent

16.8

A member of the public complains that you are making too much dust. What TWO things should you do?

- A Tell them you have nearly finished
- B Inform your supervisor immediately
- C Ignore them – they are always complaining
- D Ask your supervisor if there is an alternative way of working, such as dampening down
- E Wait until it's dark and then carry on

16.9

There has been a spillage of hydraulic oil from plant working near a watercourse. What one action should you NOT do?

- A Notify the site manager
- B Use detergents to clean up the oil
- C Contain the spillage
- D Switch the plant off

16.10

These signs tell you that a substance can be:

- A harmful
- B toxic
- C corrosive
- D harmful to the environment

16.11

What does this sign mean?

- A The spill kit has been inspected and is OK
- B Assemble here if there is a spillage
- C The equipment you need to stop, contain and clean up a spill is located here
- D A spill happened here and has been cleaned up

E
16

16.12

When a product has either of these labels, how should you dispose of it?

A. Put it in any skip or bin

B. Follow specific instructions on the label or in work instructions

C. If it is a liquid and less than one litre you can pour it down a drain

D. Leave it somewhere for other people to deal with

16.13

Who on site needs to understand relevant environmental risks on a construction site?

A. Only the principal contractor

B. Only the sub-contractors

C. All people working on site

D. Just the environmental clerk of works

16.14

Under Environmental Law, which statement is true?

A. Companies AND individuals can be prosecuted if they do not follow the law

B. It is illegal to discharge contaminated water into a watercourse

C. It is illegal to transport waste without a licence

D. All of these answers

16.15

Do you have any responsibility with regard to sustainability on site?

A. No, it is a matter for the site manager

B. No, it is a matter for the Environment Agency

C. Only on sites where there are rare species of plants

D. Yes, on every site that you go on to

16.16

Which of the following is NOT best practice from a sustainability perspective?

A. Saving materials, fuel, water and energy

B. Looking after the people working on or near the site

C. Protecting the environment

D. Sending unused and waste materials to landfill

16.17

Which of the following does NOT help sustainability on site?

A. Leaving engines, motors and other power on when not needed

B. Segregating waste

C. Lift sharing or using public transport to get to work

D. Working safely

E
16

16.18

Which of the following should you do in the interest of sustainability on site?

A Run plant and equipment when they are not needed

B Bury waste materials in the ground

C Comply with site instructions on handling waste materials

D Pour waste liquids down a drain outside the site

16.19

Which of the following is NOT part of sustainable construction?

A Creating a nuisance to the residents of neighbouring properties

B Preventing water and soil pollution

C Saving energy

D Minimising the amount of waste created in doing a job

16.20

From an environmental point of view, we should try to reuse materials because it:

A saves the client money

B takes lots of energy and 'raw' material to make most construction products

C makes less mess on site

D is a European Union law

16.21

Which TWO actions could help minimise waste?

A Reuse off-cuts (such as half bricks) as far as possible, rather than discarding them

B Use new materials/packs at the beginning of each day

C Leave bags of cement and plaster out in the rain, unprotected

D Only take or open what you need and return or reseal anything left over

E Always take much more than required – just in case you need it

16.22

Which of the following is good environmental practice?

A Over-ordering materials

B Segregating waste into different types

C Leaving skips uncovered in wet weather

D Poor storage of materials, causing pollution

E
16

16.23

Do you have any responsibility for minimising the amount of waste created?

A. Only if asbestos removal is being carried out

B. Yes, everyone on site has a responsibility to do this

C. No, it is the responsibility of the Environment Agency

D. Only during the site clean up at the end of the project

16.24

If you have unused material left, what should you always do before you consider putting it into a skip?

A. Make sure there is a label on it

B. Tell your supervisor

C. Check whether someone else can make use of it

D. Make sure there is room in the skip

16.25

Why should different types of waste be separated on site?

A. They will take up less room in the skip

B. So the Government can charge us a fair amount of Landfill Tax

C. So the client can check what is being thrown away

D. So it can be recycled more easily

E
16

Congratulations, you have now completed the core knowledge questions.

For the operatives' test

You still need to prepare for the behavioural case studies, which you can do by:

 watching the film *Setting out* at www.citb.co.uk/settingout

 reading the transcript of the film at the back of this book.

For the specialists' test

You should now revise the appropriate specialist activity from Section F.

You then need to prepare for the behavioural case studies, which you can do by:

 watching the film *Setting out* at www.citb.co.uk/settingout

 reading the transcript of the film at the back of this book.

F

Specialist activities

Contents

If you are preparing for a specialist test you also need to revise the appropriate specialist activity, from those listed below.

17.1

What is the purpose of the health and safety file on a construction project?

A To assist people who have to carry out work on the structure in the future

B To assist in the preparation of final accounts for the structure

C To record the health and safety standards of the structure

D To record the accident details

17.2

Where a project is notifiable under the current Construction (Design and Management) Regulations, what must be in place before construction work begins?

A Construction project health and safety file

B Construction phase health and safety plan

C Construction project plan

D Construction contract agreement

17.3

The current Construction (Design and Management) Regulations require a supported excavation to be inspected:

A every seven days

B at the start of every shift

C once a month

D when it is more than 2 m deep

17.4

Under the current Construction (Design and Management) Regulations, for how long must you keep inspection records on site?

A For a period of three months

B Not at all, the records need only be kept at company head offices

C Until the project is completed

D For a period of one month

17.5

Under the current Construction (Design and Management) Regulations, what document must be handed to the client upon completion of the construction phase?

A The safety log book

B The premises log book

C The site accident book

D The health and safety file

F
17

17.6

If your company is the principal contractor on a project that is notifiable under the current Construction (Design and Management) Regulations, you may come into contact with the CDM co-ordinator because of their legal duty to:

A collect accident statistics for the Health and Safety Executive (HSE)

B manage the flow of health and safety information between contractors and other parties

C supervise the principal contractor's implementation of the construction phase health and safety plan

D supervise or monitor construction work

17.7

Under the current Construction (Design and Management) Regulations, which of the following must the principal contractor ensure is specifically provided before allowing any demolition work to commence?

A A construction phase safety plan

B The arrangements for demolition recorded in writing

C A generic risk assessment

D A pre-tender health and safety plan

17.8

Apart from work for domestic clients, under the current Construction (Design and Management) Regulations, in which of the following situations must the Health and Safety Executive (HSE) be notified of a project?

A Where the work will last more than 30 days or more than 500 person-days

B Where the building and construction work will last more than 300 person-days

C When the work will take place outside normal hours

D Where there is more than one building to be erected

17.9

Under the current Construction (Design and Management) Regulations, who must develop the pre-construction information into a construction phase plan?

A CDM co-ordinator

B Designer

C Client

D Principal contractor

F
17

Answers: 17.6 = B   17.7 = B   17.8 = A   17.9 = D

17.10

A COSHH assessment tells you how:

A to lift heavy loads and how to protect yourself

B to work safely in confined spaces

C a substance might harm you and how to protect yourself when you are using it

D noise levels are assessed and how to protect your hearing

17.11

You have to use a new material for the first time and need to carry out a COSHH assessment. What are the TWO main things you will need?

A Your company's safety policy

B The material safety data sheet

C The age of the people doing the work

D The material delivery note

E Details of where, who and how you will be using the product

17.12

Which piece of equipment is used with a cable avoidance tool (CAT) to detect cables?

A Compressor

B Signal generator

C Metal detector

D Gas detector

17.13

In the colour coding of electrical power supplies on site, what voltage does a blue plug represent?

A 50 volts

B 110 volts

C 240 volts

D 415 volts

17.14

On the site electrical distribution system, which colour plug indicates a 415 volt supply?

A Yellow

B Blue

C Black

D Red

17.15

A RCD (residual current device) must be used in conjunction with 230 volt electrical equipment because it:

A lowers the voltage

B quickly cuts off the power if there is a fault

C makes the tool run at a safe speed

D saves energy and lowers costs

F
17

17.16

How could a site worker check if the RCD (residual current device) through which a 230 volt hand tool is connected to the supply is working correctly?

A   Switch the tool on and off

B   Press the test button on the RCD unit

C   Switch the power on and off

D   Run the tool at top speed to see if it cuts out

17.17

Untidy leads and extension cables are responsible for many trips and lost work time injuries. What TWO things should you do to help?

A   Run cables and leads above head height and over the top of doorways and walkways rather than across the floor

B   Tie any excess cables and leads up into the smallest coil possible

C   Keep cables and leads close to the wall and not in the middle of the floor or walkway

D   Make sure your cables go where you want them to and not worry about others

E   Unplug the nearest safety lighting and use these sockets instead

17.18

When overhead electric cables cross a construction site, it is recommended that goal-post barriers should be erected parallel to the overhead cables at a distance not less than:

A   3 m

B   4 m

C   5 m

D   6 m

17.19

An emergency route(s) must be provided on construction sites to ensure safe passage to:

A   the ground

B   open air

C   a place of safety

D   the first-aid room

17.20

Which of the following is a significant hazard when excavating alongside a building or structure?

A   Undermining or weakening the foundations of the building

B   Noise and vibration affecting the occupiers of the building

C   Excavating too deep in soft ground

D   Damage to the surface finish of the building or structure

F
17

17.21

What danger is created by excessive oxygen in a confined space?

A) Increase in breathing rate of workers

B) Increased flammability of combustible materials

C) Increased working time inside work area

D) False sense of security

17.22

When planning possible work in a confined space, what should be the first consideration?

A) How long the job will take

B) To avoid the need for operatives to enter the space

C) How many operatives will be required

D) Personal protective equipment (PPE)

17.23

Before planning for anyone to enter a confined space, following the principles of prevention, what should be the first consideration of the manager or supervisor?

A) Has the atmosphere in the confined space been tested?

B) Has a safe means of access and egress been established?

C) Is there an alternative method of doing the work?

D) Have all who intend to enter the confined space been properly trained?

17.24

When is it advisable to take precautions to prevent the fall of persons, plant or materials into an excavation?

A) At all times

B) When the excavation is 2 m or more deep

C) When the excavation is 1.2 m or more deep

D) When there is a risk from an underground cable or other service

17.25

Which of the following precautions should be taken to prevent a dumper from falling into an excavation when tipping material into it?

A) Dumpers kept 5 m away from the excavation

B) Stop blocks provided parallel to the trench appropriate to the vehicle's wheel size

C) Dumper drivers required to judge the distance carefully or given stop signals by another person

D) Cones or signage erected to indicate safe tipping point

F
17

17.26

Which TWO of the following factors must be considered when providing first-aid facilities on site?

A. The cost of first-aid equipment

B. The hazards, risks and nature of the work carried out

C. The number of people expected to be on site at any one time

D. The difficulty in finding time to purchase the necessary equipment

E. The space in the site office to store the necessary equipment

17.27

The minimum level of first-aid cover required at any workplace is an appointed person. Which of the following would you expect the appointed person to carry out?

A. Provide MOST of the care normally carried out by a first aider

B. Provide ALL of the care normally provided by a first aider

C. Contact the emergency services and direct them to the scene of an accident

D. Only apply splints to broken bones

17.28

The monitoring and controlling of health and safety procedures can be either proactive or reactive. Proactive monitoring means:

A. ensuring that staff always do the work that they have been instructed to do safely

B. deciding how to prevent accidents similar to those that have already occurred

C. looking at the work to be done, what could go wrong and how it could be done safely

D. checking that all staff read and understand all health and safety notices

17.29

Why may a young person be more at risk of having accidents?

A. Legislation does not apply to anyone under 18 years of age

B. They are usually left to work alone to gain experience

C. They have less experience and may not recognise danger or understand fully what could go wrong

D. There is no legal duty to provide them with personal protective equipment (PPE)

F
17

17.30

How should cylinders containing liquefied petroleum gas (LPG) be stored on site?

- A In a locked cellar with clear warning signs
- B In a locked, external compound at least 3 m from any oxygen cylinders
- C Within a secure storage container
- D Covered by a tarpaulin to shield the compressed cylinder from sunlight

17.31

Where should liquefied petroleum gas (LPG) cylinders be positioned when supplying an appliance in a site cabin?

- A Inside the site cabin in a locked cupboard
- B Under the cabin
- C Inside the cabin next to the appliance
- D Outside the cabin

17.32

Welding is about to start on your site. What should be used to protect passers-by from getting arc eye?

- A Warning signs
- B Screens
- C Personal protective equipment (PPE)
- D Nothing

17.33

When setting up a fuel storage tank on site, a spillage bund must have a minimum capacity of the contents of the tank, plus:

- A 10% (110% of the total content)
- B 30% (130% of the total content)
- C 50% (150% of the total content)
- D 75% (175% of the total content)

17.34

If there is a fatal accident or a reportable dangerous occurrence on site, when must the Health and Safety Executive (HSE) be informed?

- A Immediately
- B Within five days
- C Within seven days
- D Within 10 days

17.35

If a prohibition notice is issued by an inspector of the Health and Safety Executive (HSE) or local authority:

- A work can continue, provided that a risk assessment is carried out
- B the work that is subject to the notice must cease
- C the work can continue if extra safety precautions are taken
- D the work in hand can be completed, but no new works started

F
17

17.36

Who should you inform if someone reports to you that they have work-related hand-arm vibration syndrome?

A The Health and Safety Executive (HSE)

B The local Health Authority

C Their doctor

D The nearest hospital

17.37

An employer has to prepare a written health and safety policy if:

A they employ five people or more

B they employ three people or more

C they employ a safety officer

D the work is going to last more than 30 days

17.38

The significant findings of risk assessments must be recorded when more than a certain number of people are employed. How many?

A Three or more

B Five or more

C Six or more

D Seven or more

17.39

Before allowing a lifting operation to be carried out, you must ensure that the sequence of operations to enable a lift to be carried out safely is confirmed in:

A the crane hire contract

B an approved lifting plan or method statement

C a lifting operation toolbox talk

D a risk assessment

17.40

What does the term 'lower exposure action value' mean when referring to noise?

A The average background noise level

B The noise level at which the worker can request hearing protection

C The level of noise which must not be exceeded on the site boundary which causes noise nuisance

D The noise level at which the worker must wear hearing protection

17.41

At what decibel (dB(A)) level does it become mandatory for an employer to establish hearing protection zones?

A 80 dB(A)

B 85 dB(A)

C 90 dB(A)

D 95 dB(A)

F
17

17.42

At what minimum noise level must you provide hearing protection to workers if they ask for it?

- A 80 decibels
- B 85 decibels
- C 87 decibels
- D 90 decibels

17.43

The significance of a weekly or daily personal noise exposure of 87 decibels dB(A) is that:

- A it is the lower action value and no action is necessary
- B it is the upper action value and hearing protection must be issued
- C it is the peak sound pressure and all work must stop
- D it is the exposure limit value and must not be exceeded

17.44

In considering what measures to take to protect people against risks to their health and safety, personal protective equipment (PPE) should always be regarded as:

- A the first line of defence
- B the only practical measure
- C the best way to tackle the job
- D the last resort

17.45

In deciding what control measures to take, following a risk assessment that has revealed a risk, what measure should you always consider first?

- A Make sure personal protective equipment (PPE) is available
- B Adapt the work to the individual
- C Give priority to measures that protect the whole workforce
- D Avoid the risk altogether if possible

17.46

Why is it important that hazards are identified?

- A They have the potential to cause injury or damage
- B They must all be eliminated before work can start
- C They must all be notified to the Health and Safety Executive (HSE)
- D So toolbox talks can be given on the hazards

17.47

In the context of a risk assessment, what does the term 'risk' mean?

- A Something with the potential to cause injury
- B An unsafe act or condition
- C The likelihood or chance that a hazard could actually cause harm or damage
- D Any work activity that can be described as dangerous

F
17

17.48

What must a sub-contractor provide you with in relation to a worker who is 17 years old?

A A mentor or 'buddy' to stay with them at all times

B Health and Safety Executive (HSE) permission for the 17-year-old to be on site

C The legal guardian's permission for the 17-year-old to be on site

D A risk assessment addressing the issue of young persons

17.49

The number of people who may be carried in a passenger hoist on site must be:

A displayed on a legible notice in the site welfare area

B displayed on a legible notice within the cage of the hoist

C explained in the site induction

D explained to the hoist operator

17.50

From a safety point of view, which of the following should be considered first when deciding on the number and location of access and egress points to a site?

A Off road parking for cars and vans

B Access for the emergency services

C Access for heavy vehicles

D Site security

17.51

How should access be controlled, if people are working in a riser shaft?

A By a site security operative

B By those who are working in it

C By the main contractor

D By a permit to work system

17.52

What is the purpose of using a 'permit to work' system?

A To ensure that the job is being carried out properly

B To ensure that the job is carried out by the easiest method

C To enable tools and equipment to be properly checked before work starts

D To establish a safe system of work

17.53

Employers must prevent exposure of their workers to substances hazardous to health, where this is reasonably practicable. If it is not reasonably practicable, which of the following should be considered first?

A What instruction, training and supervision to provide

B What health surveillance arrangements will be needed

C How to minimise risk and control exposure

D How to monitor the exposure of workers in the workplace

F
17

17.54

If a scaffold is not complete, which of the following actions should be taken by the site manager?

A Make sure the scaffolders complete the scaffold

B Tell all operatives not to use the scaffold

C Use the scaffold with care and display a warning notice

D Prevent access to the scaffold by unauthorised people and add warning signage and barriers where required

17.55

Following a scaffold inspection under the Work at Height Regulations, how soon must a report be given to the person on whose behalf the inspection was made?

A Within two hours

B Within six hours

C Within 12 hours

D Within 24 hours

17.56

On a scaffold the minimum height of the main guard-rail must be:

A 875 mm

B 910 mm

C 950 mm

D 1,000 mm

17.57

On a scaffold the unprotected gap between any guard-rail, toe-board, barrier or other similar means of protection should NOT exceed:

A 400 mm

B 470 mm

C 500 mm

D 600 mm

17.58

What is the best way for a supervisor or manager to make sure that the operatives doing a job have fully understood a method statement?

A Put the method statement in a labelled spring-binder in the office

B Explain the method statement to those doing the job and test their understanding

C Make sure that those doing the job have read the method statement

D Display the method statement on a noticeboard in the office

F
17

17.59

What is your LEAST reliable source of information when assessing the level of vibration from a powered, percussive hand tool?

A In-use vibration measurement of the tool

B Vibration figures taken from the tool manufacturer's handbook

C Your own judgement based upon observation or experience

D Vibration data from the Health and Safety Executive's (HSE) master list

17.60

You will find the details of the welfare facilities that must be provided on site in which regulations?

A The Construction (Health, Safety and Welfare) Regulations

B The Construction (Design and Management) Regulations

C The Management of Health and Safety at Work Regulations

D The Workplace (Health, Safety and Welfare) Regulations

17.61

What is regarded as the last resort in the hierarchy of control for operatives' safety when working at height?

A Safety harness

B Mobile elevating work platform (MEWP)

C Mobile tower scaffold

D Access tower scaffold

17.62

Which of the following is a fall-arrest system?

A Guard-rail and toe-board

B Scaffold towers

C Mobile elevating work platform (MEWP)

D Safety harness and lanyard

17.63

Under the requirements of the Work at Height Regulations, the minimum width of a working platform must be:

A suitable and sufficient for the job in hand

B two scaffold boards wide

C three scaffold boards wide

D four scaffold boards wide

F
17

17.64

The Work at Height Regulations require a working platform to be inspected by a competent person:

A after an accident

B every day

C fortnightly

D before first use and then every seven days afterwards

17.65

For a ladder, what is the maximum vertical height that may be climbed before an intermediate landing place is required?

A 7.5 m

B 8 m

C 8.5 m

D 9 m

17.66

The advantage of using safety nets rather than harness and lanyard is that:

A safety nets do not need inspecting

B workers' lanyards can get entangled with other workers' lanyards

C safety nets provide collective fall protection

D safety nets can be rigged by anyone

17.67

What should you do if you notice that operatives working above a safety net are dropping off-cuts of material and other debris into the net?

A Nothing, as at least it is all collecting in one place

B Ensure that the net is cleared of debris weekly

C Have the net cleared and ensure it is not allowed to happen again

D Ensure that the net is cleared of debris daily

17.68

What should be included in a safety method statement for working at height? Give THREE answers.

A The cost of the job and time it will take

B The sequence of operations and the equipment to be used

C How much insurance cover will be required

D How falls are to be prevented

E Who will supervise the job on site

F
17

17.69

When putting people to work above public areas, your first consideration should be to:

A minimise the number of people below at any one time

B prevent complaints from the public

C let the public know what you are doing

D prevent anything falling on to people below

17.70

A competent person must routinely inspect a working platform:

A after it is erected and at intervals not exceeding seven days

B only after it has been erected

C after it is erected and then at monthly intervals

D after it is erected and then at intervals not exceeding 10 days

17.71

Ideally, a safety net should be rigged:

A immediately below where you are working

B 2 m below where you are working

C 6 m below where you are working

D at any height below the working position

17.72

What is the MAIN danger of leaving someone who has fallen suspended in a harness for too long?

A The anchorage point may fail

B They may try to climb back up the structure and fall again

C They may suffer severe trauma or even death

D It is a distraction for other workers

17.73

Edge protection must be designed to:

A allow persons to work both sides

B secure tools and materials close to the edge

C warn people where the edge of the roof is

D prevent people and materials falling

17.74

When should guard-rails be fitted to a working platform?

A If it is possible to fall 2 m

B At any height if a fall could result in an injury

C If it is possible to fall 3 m

D Only if materials are being stored on the working platform

F
17

Answers: 17.69 = D   17.70 = A   17.71 = A   17.72 = C   17.73 = D   17.74 = B   **109**

17.75

The Beaufort Scale is important when working at height externally because it measures the:

A ratio of sloping ground to height

B load-bearing capacity of a flat roof

C wind speed

D load-bearing capacity of a scaffold

17.76

A design feature of some airbags used for fall arrest is a controlled leak rate. If you are using these, the inflation pump must:

A be electrically powered

B be switched off from time to time to avoid over-inflation

C run all the time while work is carried out at height

D be switched off when the airbags are full

17.77

Why is it dangerous to use inflatable airbags for fall arrest that are too big for the area to be protected?

A They will exert a sideways pressure on anything that is containing them

B The pressure in the bags will cause them to burst

C The inflation pump will become overloaded

D They will not fully inflate

F
17

18.1

If asbestos is present what should happen before demolition or refurbishment takes place?

A Advise workers that asbestos is present and continue with demolition

B All asbestos should be removed as far as reasonably practicable

C Advise the Health and Safety Executive (HSE) that asbestos is present and continue with demolition

D Inspect the condition of the asbestos materials

18.2

What kind of survey is required to identify asbestos prior to demolition?

A Type 3 survey

B Management survey

C Refurbishment and demolition survey

D Type 2 survey

18.3

Every demolition contractor undertaking demolition operations must first appoint:

A a competent person to supervise the work

B a sub-contractor to strip out the buildings

C a safety officer to check on health and safety compliance

D a quantity surveyor to price the extras

18.4

If there are any doubts as to a building's stability, a demolition contractor should consult:

A another demolition contractor

B a structural engineer

C a Health and Safety Executive (HSE) factory inspector

D the company safety adviser

18.5

Which of the following pieces of equipment can a 17-year-old trainee demolition operative use unsupervised?

A Excavator 360°

B Dump truck

C Wheelbarrow

D Rough terrain forklift

18.6

The operation of a scissor lift would be unsafe if:

A the controls on the platform are used

B the ground is soft and sloping

C weather protection is not fitted

D the machine is short of fuel

F
18

18.7

On site, what is the minimum distance that oxygen should be stored away from propane, butane or other gases?

- (A) 1 m
- (B) 2 m
- (C) 3 m
- (D) 4 m

18.8

Where should liquefied petroleum gas (LPG) cylinders be located when being used for heating or cooking in site cabins?

- (A) Under the kitchen area work surface
- (B) Inside but near the door for ventilation
- (C) In a nearby storage container
- (D) Outside the cabin

18.9

What type of fire extinguisher should NOT be provided where petrol or diesel is being stored?

- (A) Foam
- (B) Water
- (C) Dry powder
- (D) Carbon dioxide

18.10

Continual use of hand-held breakers or drills is most likely to cause:

- (A) dermatitis
- (B) Weil's disease (leptospirosis)
- (C) vibration white finger
- (D) skin cancer

18.11

What is the most common source of high levels of lead in the blood of operatives doing demolition work?

- (A) Stripping lead sheeting
- (B) Cutting lead-covered cable
- (C) Cold cutting fuel tanks
- (D) Hot cutting coated steel

18.12

Which of the following items of personal protective equipment (PPE) provides the lowest level of protection when working in dusty conditions?

- (A) Half-mask dust respirator
- (B) Positive pressure-powered respirator
- (C) Compressed airline breathing apparatus
- (D) Self-contained breathing apparatus

F
18

18.13

Which TWO of the following would be suitable to use when cutting coated steelwork?

A A disposable dust mask

B Positive pressure-powered respirator

C High-efficiency dust respirator

D Ventilated helmet respirator

E A nuisance dust mask

18.14

After exposure to lead, what precautions should you take before eating or drinking?

A Wash your hands and face

B Do not smoke

C Change out of dirty clothes

D Rinse your mouth with clean water

18.15

Where do you find information about daily checks required for mobile plant?

A On the stickers attached to the machine

B In the manufacturer's handbook

C In the supplier's information

D All of these answers

18.16

What should you do while reversing mobile plant if you lose sight of the signaller or slinger who is directing you?

A Carry on reversing slowly

B Stop the vehicle

C Adjust your wing mirror

D Sound the horn and move forward

18.17

How often should lifting equipment, which is not used to lift people, be thoroughly examined?

A Once every six months

B Every two years

C A minimum of once a year

D Every 14 months

18.18

When leaving mobile plant unattended you should:

A leave the engine running, if safe to do so

B park it in a safe place, remove the keys and lock it

C put the parking brake on and tell people not to use it

D put a sign saying 'No unauthorised access' on it

F
18

18.19

Which statement is true with regard to using machines?

A) Guards can be removed to make work easier

B) It's OK to wear rings and other jewellery as long as you take care

C) Carefully remove waste material while the machine is in motion

D) Never use a machine unless you have been trained and given permission to use it

18.20

Which of the following is NOT a part of a plant operator's daily pre-use check?

A) Emergency systems

B) Engine oil level

C) Hydraulic fluid level

D) Brake pad wear

18.21

An operator of a scissor lift must:

A) be trained and authorised in the use of the equipment

B) only use the ground level controls

C) be in charge of the work team

D) ensure that only one person is on the platform at any time

18.22

On demolition sites, what must the drivers of plant have, for their own and others' safety?

A) Adequate visibility from the driving position

B) A temperature controlled cab

C) Wet weather gear when it's raining

D) A supervisor in the cab with them

18.23

When must head and tail lights be used on mobile plant?

A) Only if using the same traffic route as private cars

B) In all conditions of poor visibility

C) When operated by a trainee

D) Only if crossing pedestrian routes

18.24

With regard to mobile plant, what safety feature is provided by FOPS?

A) The speed is limited when tracking over hard surfaces

B) The machine stops automatically if the operator lets go of the controls

C) The operator is protected from falling objects

D) The reach is limited when working near to live overhead cables

F
18

Answers: 18.19 = D   18.20 = D   18.21 = A   18.22 = A   18.23 = B   18.24 = C

18.25

What should you do if you discover underground services not previously identified?

[A] Fill in the hole and say nothing to anyone

[B] Stop work until the situation has been resolved

[C] Cut the pipe or cable to see if it's live

[D] Get the machine driver to dig it out

18.26

What action should you take if you discover unlabelled drums or containers on site?

[A] Put them in the nearest waste skip

[B] Ignore them. They will get flattened during the demolition

[C] Stop work until they have been safely dealt with

[D] Open them and smell the contents

18.27

The plant you are driving has defective brakes. What action should you take?

[A] Reduce your speed accordingly

[B] Report it and carry on working

[C] Report it and isolate the machine

[D] Use the handbrake until the machine is fixed

18.28

What action should be taken if a wire rope sling is defective?

[A] Do not use it and make sure that no-one else can

[B] Only use it for up to half its safe working load

[C] Put it to one side to wait for repair

[D] Only use it for small lifts under 1 tonne

18.29

Which TWO of the following documents refer to the specific hazards associated with demolition work in confined spaces?

[A] Safety policy

[B] Permit to work

[C] Risk assessment

[D] Scaffolding permit

[E] Hot work permit

18.30

When asbestos material is suspected in buildings to be demolished, what is the FIRST priority?

[A] A competent person carries out an asbestos survey

[B] Notify the Health and Safety Executive (HSE) of the possible presence of asbestos

[C] Remove and dispose of the asbestos

[D] Employ a licensed asbestos remover

F
18

18.31

What is the safest method of demolishing brick or internal walls by hand?

A Undercut the wall at ground level

B Work across in even courses from the ceiling down

C Work from the doorway at full height

D Cut down at corners and collapse in sections

18.32

Who should be consulted before demolition is carried out near to overhead cables?

A The Health and Safety Executive (HSE)

B The fire service

C The electricity supply company

D The land owner

18.33

When demolishing a building in controlled sections, what is the most important consideration for the remaining structure?

A The soft strip is completed

B All non-ferrous metals are removed

C It remains stable

D Trespassers cannot get in at night

18.34

Where would you find the intended method of controlling identified hazards on a demolition project?

A The demolition toolbox

B The demolition plan

C The pre-tender health and safety plan

D The construction phase health and safety plan

18.35

How must cans or drums be stored, to prevent any leakage spreading?

A On wooden pallets off the ground

B On their sides and chocked to prevent movement

C Upside down to prevent water penetrating the screw top

D In a bund in case of leakage

18.36

Before entering large, open-topped tanks, what is the most important thing you should obtain?

A A ladder for easy access

B A valid permit to work

C An operative to keep watch over you

D A gas meter to detect any gas

F
18

18.37

When hinge-cutting a steel building or structure for a 'controlled collapse', which should be the last cuts?

A. Front leading row top cuts

B. Front leading row bottom cuts

C. Back row top cuts

D. Back row bottom cuts

18.38

What safety devices should be fitted between the pipes and the gauges of oxy-propane cutting equipment?

A. Non-return valves

B. On-off taps

C. Flame retardant tape

D. Flashback arresters

18.39

What type of fire extinguisher should you NOT use in confined spaces?

A. Water

B. Carbon dioxide

C. Dry powder

D. Foam

18.40

Before carrying out the demolition cutting of fuel tanks what should be obtained?

A. A gas free certificate

B. An isolation certificate

C. A risk assessment

D. A COSHH assessment

18.41

How long is a gas free certificate issued for?

A. One week

B. One month

C. One day

D. One hour

18.42

What do the letters SWL stand for?

A. Satisfactory working limit

B. Safe working level

C. Satisfactory weight limit

D. Safe working load

18.43

Which of the following is true as regards the safe working load of a piece of equipment?

A. It must never be exceeded

B. It is a guide figure that may be exceeded slightly

C. It may be exceeded by 10% only

D. It gives half the maximum weight to be lifted

18.44

What should be clearly marked on all lifting gear?

A. Date of manufacture

B. Name of maker

C. Date next test is due

D. Safe working load

F
18

18.45

Lifting accessories must be thoroughly examined every:

A three months

B six months

C 14 months

D 18 months

18.46

Plant and equipment needs to be inspected and the details recorded by operators:

A daily at the beginning of each shift

B weekly

C monthly

D every three months

18.47

What is the importance of having ROPS fitted to some mobile plant?

A It ensures that the tyre pressure is correct

B It protects the operator if the machine rolls over

C It prevents over-pressurisation of the hydraulic system

D It prevents unauthorised passengers being carried

18.48

The correct way to climb off a machine is to:

A jump down from the seated position

B climb down, facing forward

C climb down, facing the machine

D use a ladder

18.49

It is acceptable to carry passengers on a machine provided:

A the employer gives permission

B they are carried in the skip

C the machine is fitted with a purpose-made passenger seat

D the maximum speed is no greater than 10 mph

18.50

With regard to the safe method of working, what is the most important subject of induction training for demolition operatives?

A Working hours on the site

B Explanation of the method statement

C Location of welfare facilities

D COSHH assessments

Answers: 18.45 = B   18.46 = B   18.47 = B   18.48 = C   18.49 = C   18.50 = B

F
18

19.1

While climbing a ladder, a colleague slips and falls about 2 m. They are lying on the ground saying that their back hurts. What is the first thing that you should do?

A Go and put it in the accident book

B Help them get up

C Tell them to lie still and send someone else to get a first aider

D Make sure there is nothing wrong with the ladder

19.2

When using a blowtorch near to flexible pipe lagging, you should:

A just remove enough lagging to carry out the work

B remove the lagging at least 1 m either side of the work

C remove the lagging at least 3 m either side of the work

D wet the lagging but leave it in place

19.3

You notice that the handle of your wooden 'rat tail' file is starting to split. What should you do?

A Wrap it tightly with plastic adhesive tape

B Use it until the handle falls off completely and then replace the handle

C Remove the damaged handle and replace it

D Throw the damaged tool away

19.4

What is the most likely risk of injury when cutting large diameter pipe?

A Your fingers may become trapped between the cutting wheel and the pipe

B The inside edge of the cut pipe becomes extremely sharp to touch

C Continued use can cause muscle damage

D Pieces of sharp metal could fly off

19.5

You have removed a WC pan in a public toilet and notice a hypodermic syringe lodged in the soil pipe connector. What should you do?

A Ensure the syringe is empty, remove the syringe and place it with the rubbish

B Wear gloves, break the syringe into small pieces and flush it down the drain

C Notify the supervisor, cordon the area off and call the emergency services

D Wearing gloves, use grips to remove the syringe to a safe place and tell your supervisor

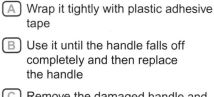

F
19

19.6

You have been handling sheet lead. How is some lead most likely to get in your bloodstream?

A By not using the correct respirator

B By not washing your hands before eating

C By not changing out of your work clothes

D By not wearing safety goggles

19.7

The legionella bacteria that cause legionnaires' disease are most likely to be found in which of the following?

A A boiler operating at a temperature of 80°C

B A shower hose outlet

C A cold water storage cistern containing water at 10°C

D A WC toilet pan

19.8

How are legionella bacteria passed on to humans?

A Through fine water droplets, such as sprays or mists

B By drinking dirty water

C Through contact with the skin

D From other people when they sneeze

19.9

When bossing a sheet lead corner using lead-working tools, you:

A are allowed to smoke during the bossing process

B should not smoke during the bossing process

C should only smoke when your hands are protected with barrier cream

D are allowed to smoke if you are wearing gloves

19.10

If breaking up a cast iron bath, which of the following is the proper way to protect your hearing?

A A portable stereo and head set

B Cotton wool pads

C Ear defenders

D Rolled-up tissue paper

19.11

What item of personal protective equipment (PPE), from the following list, should be used when oxyacetylene welding?

A Ear defenders

B Clear goggles

C Green-tinted goggles

D Dust mask

F
19

Answers: 19.6 = B   19.7 = B   19.8 = A   19.9 = B   19.10 = C   19.11 = C

19.12

You are drilling a hole through a metal partition to receive a 15 mm pipe from a radiator. You need to wear eye protection:

A when drilling overhead only

B when the drill bit exceeds 20 mm

C always, whatever the circumstances

D when drilling through concrete only

19.13

In plumbing work, what part of the body could suffer long-term damage when hand bending copper pipe using an internal spring?

A Elbows

B Hands

C Back

D Knees

19.14

When repairing a burst water main using pipe-freezing equipment to isolate the damaged section of pipe, you should:

A always work in pairs when using pipe-freezing equipment

B never allow the freezing gas to come into direct contact with surface water

C never use pipe-freezing equipment on plastic pipe

D wear gloves to avoid direct contact with the skin

19.15

You are drilling a 100 mm diameter hole for a flue pipe through a brick wall with a large hammer drill. Which combination of personal protective equipment (PPE) should you be supplied with?

A Gloves, breathing apparatus and boots

B Ear defenders, face mask and boots

C Ear defenders, breathing apparatus and barrier cream

D Barrier cream, boots and face mask

19.16

When working with fibreglass roof insulation, which of the following items of personal protective equipment (PPE) should you wear?

A Gloves, face mask and eye protection

B Boots, eye protection and ear defenders

C Ear defenders, face mask and boots

D Barrier cream, eye protection and face mask

19.17

Where might you come across asbestos?

A Insulation board around radiators

B Gaskets and seals on joints

C Rope seals on a boiler

D All of these answers

F
19

19.18

While working, you come across a hard, white, powdery material that could be asbestos. What should you do?

A While wearing a face mask, remove the material and dispose of it safely

B Remove the material, putting it back after finishing the job

C Stop work immediately and tell your supervisor about the material

D Dampen the material down with water and remove it before carrying out the work

19.19

Why is it important that operatives know the difference between propane and butane equipment?

A Propane equipment operates at higher pressure

B Propane equipment operates at lower pressure

C Propane equipment is cheaper

D Propane equipment can be used with smaller, easy-to-handle cylinders

19.20

Which of the following statements is true?

A Both propane and butane are heavier than air

B Butane is heavier than air while propane is lighter than air

C Propane is heavier than air while butane is lighter than air

D Both propane and butane are lighter than air

19.21

Apart from the cylinders used in gas-powered forklift trucks, you should never see liquefied petroleum gas (LPG) cylinders placed on their sides during use because:

A it would give a faulty reading on the contents gauge, resulting in flashback

B air could be drawn into the cylinder, creating a dangerous mixture of gases

C the liquid gas would be at too low a level to allow the torch to burn correctly

D the liquid gas could be drawn from the cylinder, creating a safety hazard

19.22

What is the preferred method of checking for leaks when assembling liquefied petroleum gas (LPG) equipment before use?

A Test with a lighted match

B Sniff the connections to detect the smell of gas

C Listen to hear for escaping gas

D Apply leak detection fluid to the connections

F
19

19.23

When transporting liquefied petroleum gas (LPG) cylinders (above 5 kg) in an enclosed van, under the Packaged Goods Regulations the driver must:

A be trained and competent in hazards relating to LPG

B have a heavy goods vehicle licence

C under no circumstances carry LPG cylinders

D have a full driving licence with a Packaged Goods Regulations endorsement

19.24

What is the colour of propane gas cylinders?

A Black

B Maroon

C Red/orange

D Blue

19.25

The use of oxyacetylene equipment is NOT recommended for which of the following jointing methods?

A Jointing copper pipe using hard soldering

B Jointing copper tube using capillary soldered fittings

C Jointing mild steel tube

D Jointing sheet lead

19.26

Which of the following makes it essential to take great care when handling oxygen cylinders?

A They contain highly flammable compressed gas

B They contain highly flammable liquid gas

C They are filled to extremely high pressures

D They contain poisonous gas

19.27

Which of the following is the safest place to store oxyacetylene gas-welding bottles when they are not in use?

A Outside in a special storage compound

B In company vehicles

C Inside the building in a locked cupboard

D In the immediate work area, ready for use the next day

19.28

When using a blowtorch to joint copper tube and fittings in a domestic property, a fire extinguisher should be:

A available in the immediate work area

B held over the joint while you are using the blowtorch

C used to cool the fitting

D available only if a property is occupied

F
19

Answers: 19.23 = A   19.24 = C   19.25 = B   19.26 = C   19.27 = A   19.28 = A

19.29

When using a blowtorch, you should:

A. stop using the blowtorch immediately before leaving the job

B. stop using the blowtorch at least one hour before leaving the job

C. stop using the blowtorch at least two hours before leaving the job

D. stop using the blowtorch at least four hours before leaving the job

19.30

When using a blowtorch near to timber, you should:

A. carry out the work taking care not to catch the timber

B. use a non-combustible mat and have a fire extinguisher ready

C. wet the timber first and have a bucket of water handy

D. point the flame away from the timber and have a bucket of sand ready to put out the fire

19.31

What is the colour of an acetylene cylinder?

A. Orange

B. Black

C. Green

D. Maroon

19.32

When using oxyacetylene welding equipment, the bottles should be:

A. laid on their side

B. stood upright

C. stood upside down

D. angled at 45°

19.33

You are asked to join plastic soil pipes in the roof space of a building using a strong-smelling hazardous solvent, but you have not been provided with any respiratory protective equipment (RPE). What should you do?

A. Just get on and do the job

B. Use a dust mask

C. Sniff the solvent to see if it has any ill-effects on you

D. Stop what you are doing and get out

19.34

You are required to replace below-ground drainage pipework in an excavation, which is approximately 2.5 m deep. The trench sides show signs of collapse. What do you do?

A. Get on with the work as quickly as possible

B. Refuse to do the work until the trench sides have been properly shored

C. Get a mate to help you so that they can pass the materials down to you

D. Ensure that you do the work with a rope around you so that you can be pulled out

F
19

19.35

You are working in an occupied building and have taken up six lengths of 3 m floor boarding when you are called away to an urgent job. What should you do?

A Leave the job as it is

B Cordon off the work area before leaving the job

C Permanently re-fix the floorboards and floor coverings

D Tell other workers to be careful while you are away

19.36

You are preparing to use an electric-powered threading machine. Which of the following statements should apply?

A The power supply should be 24 volts and the machine fitted with a guard

B The power supply should be 415 volts and the machine fitted with a guard

C Ensure your clothing cannot get caught on rotating parts of the machine

D Ear defenders should be available and should be in good condition

19.37

In terms of accident prevention, which TWO of the following are the most important precautions when working in the roof space of an occupied house?

A Safe foot access over the ceiling joists

B Safe access into the roof space using a stepladder or ladder

C Check that the roof space is free from hornets' nests

D The roof space must be fitted with a light socket and switch

E Carry out a test to see if the plasterboard ceiling will support your weight

19.38

When replacing an electrical immersion heater in a hot water storage cylinder, what should you do to make sure that the electrical supply is dead before starting plumbing work?

A Switch off and disconnect the supply to the immersion heater

B Switch off and cut through the electric cable with insulated pliers

C Switch off and test the circuit

D Switch off, isolate the supply at the mains board and test the circuit

F
19

19.39

The reason for carrying out temporary continuity bonding before removing and replacing sections of metallic pipework is to:

A provide a continuous earth for the pipework installation

B prevent any chance of blowing a fuse

C maintain the live supply to the electrical circuit

D prevent any chance of corrosion to the pipework

19.40

You are required to re-fix a section of external rainwater pipe using a power drill in wet weather conditions. Which type of drill is most suitable?

A Battery-powered drill

B Drill with 110 volt power supply

C Drill with 240 volt power supply

D Any mains voltage drill with a power breaker

19.41

What piece of equipment would you use to find out whether a section of solid wall that you are about to drill into contains electric cables?

A A neon screwdriver

B A cable tracer

C A multimeter

D A hammer and chisel

19.42

When maintaining or installing a central heating pump in a domestic property, the correct electricity supply to the pump should be:

A 24 volts

B 110 volts

C 240 volts

D 415 volts

19.43

Which of the following statements is true? It is safe to transport workers to the workplace in the rear of a van if the:

A driver has a heavy goods vehicle licence

B van is fitted with temporary seating

C van is fitted with proper seating and seatbelts

D driver has a public service vehicle licence

19.44

Which is the safest method of taking long lengths of copper pipe by van?

A Tying the pipes to the roof with copper wire

B Someone holding the pipes on the roof rack as you drive along

C Putting the pipes inside the van with the ends out of the passenger window

D Using a pipe rack fixed to the roof of the van

F
19

19.45

When should you wear a safety belt when driving a dumper truck on site?

A When travelling on rough ground

B When the dumper is loaded

C When carrying passengers

D Whenever one is provided

19.46

If lifting a roll of Code 5 sheet lead, what is the first thing you should do?

A Weigh the roll of lead

B Have a trial lift to see how heavy it feels

C Assess the whole task

D Ask your workmate to give you a hand

19.47

You need to move a cast iron bath that is too heavy to lift. You should:

A inform your supervisor and ask for assistance

B get some lifting tackle

C give it another try

D try and find someone to give you a hand

19.48

When installing a flue liner in an existing chimney, you should:

A insert the liner from ground level, pushing it up the chimney

B work in pairs and insert the liner from roof level, working off the roof

C work in pairs and insert the liner from roof level, working from roof ladders or a chimney scaffold

D break through the chimney in the loft area and insert the liner from there

19.49

You are asked to move a cast iron boiler some distance. What should you do?

A Get a workmate to carry it

B Drag it

C Roll it end-over-end

D Use a suitable trolley or other manual handling aid

19.50

If working on a plumbing job where noise levels are rather high, who would you expect to carry out noise assessment?

A A fully qualified plumber

B Your supervisor

C The site engineer

D A competent person

F
19

Answers: 19.45 = D   19.46 = C   19.47 = A   19.48 = C   19.49 = D   19.50 = D

19.51

During a refurbishment job, you may need to work under a ground-level suspended timber floor. What is the first question you should ask?

A Can the work be performed from outside?

B Will temporary lighting be used?

C What is contained under the floor?

D How many ways in or out are there?

19.52

You are fixing sheet lead flashing to a chimney on the roof of a busy town centre shop. What is the most important thing you must do?

A Pull the ladder onto the roof to prevent the public from climbing up

B Make provision for protecting the public from objects that could fall

C Wear a face mask to protect you from breathing the chimney fumes

D Wear safety boots to prevent you from slipping off the roof

19.53

Stepladders must only be used:

A inside buildings

B if they are in good condition and suitable

C if they are made of aluminium

D if they are less than 1.75 m high

19.54

You are removing guttering from a large, single-storey, metal-framed and cladded building. The job is likely to take all day. What is the most appropriate type of access equipment you could use?

A A ladder

B A mobile tower scaffold

C A putlog scaffold

D A trestle scaffold

19.55

You arrive at a job that involves using ladder access to the roof. You notice the ladder has been painted. You should:

A only use the ladder if it is made of metal

B only use the ladder if it is made of wood

C only use the ladder if wearing rubber-soled boots to prevent slipping

D not use the ladder, and report the matter to your supervisor

F
19

19.56

When dismantling lengths of cast iron soil pipe at height, you should:

A work in pairs, break the length at the collar and remove the pipe section

B tie a rope to the pipe and pull the section down. Shout 'below' to warn others

C crack the pipe at the joint and push it away, making sure the area is clear

D smash the pipe into sections using a hammer and remove it piece by piece

F
19

20.1

From a safety point of view, diesel must not be used to prevent asphalt sticking to the bed of lorries because:

- Ait will create a slipping hazard
- B it will corrode the bed of the lorry
- C it will create a fire hazard
- D it will react with the asphalt, creating explosive fumes

20.2

You are moving or laying slabs, paving blocks or kerbs on site. Which of the following TWO methods would be classed as manual handling?

- A Using a 'scissor lifter attachment' on an excavator
- B Using a trolley or sack barrow
- C Using a 'suction lifter' attached to an excavator
- D Using a two man 'scissor lifter'
- E Using a crane with fork attachment

20.3

What are TWO effects of under-inflated tyres on the operation of a machine?

- A It decreases the operating speed of the engine
- B It leads to instability of the machine
- C It causes increased tyre wear
- D It decreases tyre wear
- E It increases the operating speed of the engine

20.4

If you are driving any plant that may overturn, when must a seatbelt be worn?

- A When travelling over rough ground
- B When the vehicle is loaded
- C When you are carrying passengers
- D At all times

20.5

What precautions should be taken when parking on a gradient with the front of the vehicle pointing down the hill?

- A Engine off and turn wheels onto full lock
- B Handbrake on, engine off and remove the key
- C Leave the engine running and park crosswise
- D Apply the handbrake and put chocks under the wheels

20.6

If the dead man's handle on a machine does not operate, what should you do?

- A Keep quiet in case you get the blame
- B Report it at the end of the shift
- C Try and fix it or repair it yourself
- D Stop and report it immediately to your supervisor

F
20

Answers: 20.1 = A   20.2 = B, D   20.3 = B, C   20.4 = D   20.5 = B   20.6 = D

20.7

You may carry passengers on vehicles:

A if your supervisor gives you permission

B only if a suitable secure seat is provided for each of them

C only when off the public highway

D if you have a full driving licence

20.8

When tipping into an excavation is necessary, what is the preferred method of preventing the vehicle getting too close to the edge?

A Signage

B Driver's experience

C Signaller

D Stop-blocks

20.9

Which of these is the safest method of operating a machine that has an extending jib or boom under electrical power lines?

A Erect signs to warn drivers while they are operating the machines

B Adapt the machines to limit the extension of the jib or boom

C Place a warning notice on the machine

D Let down the tyres on the machine to increase the clearance

20.10

When using lifting equipment, such as a cherry picker, lorry loader or excavator, what does the indicated safe working load mean?

A It must never be exceeded

B It is a guide figure that may be exceeded slightly

C It may be exceeded by 10% only

D It gives half the maximum weight to be lifted

20.11

Which of the following checks should the operator of a mobile elevating work platform (MEWP), for example a cherry picker, carry out before using it?

A Check that a seatbelt is provided for the operator

B Check that a roll-over cage is fitted

C Drain the hydraulic system

D Check that emergency systems operate correctly

20.12

Mobile highways works are being carried out by day. A single vehicle is being used. What must be conspicuously displayed on or at the rear of the vehicle?

A Road narrows (left or right)

B A specific task warning sign (for example, gully cleaning)

C A 'keep left/right' arrow

D A 'roadworks ahead' sign

F
20

20.13

When should you switch on the amber flashing beacon fitted to your highways vehicle?

A) At all times

B) When travelling to and from the depot

C) When the vehicle is being used as a works vehicle

D) Only in poor visibility

20.14

When driving into a site works access on a motorway what must you do approximately 200 m before the access?

A) Switch on the vehicle hazard lights

B) Switch on the flashing amber beacon

C) Switch on the headlights

D) Switch on the flashing amber beacon and the appropriate indicator

20.15

Lifting equipment for carrying persons, for example a cherry picker, must be thoroughly examined by a competent person every:

A) 6 months

B) 12 months

C) 18 months

D) 24 months

20.16

When towing a trailer fitted with independent brakes, what must you do?

A) Fit a safety chain

B) Fit a cable that applies the trailer's brakes if the tow hitch fails

C) Use rope to secure the trailer to the tow hitch

D) Drive at a maximum speed of 25 mph

20.17

When getting off plant and vehicles, what must you do?

A) Look before you jump

B) Use the wheels and tyres for access

C) Maintain three points of contact with the vehicle

D) Jump down facing the vehicle

20.18

When leaving plant unattended what should you do?

A) Leave the amber flashing beacon on

B) Apply brake, switch off engine and remove the key

C) Leave it in a safe place with the engine ticking over

D) Park with blocks under the front wheels

F
20

20.19

What must you have before towing a compressor on the highway? Give TWO answers.

- A Permission from your supervisor
- B The correct class of driving licence
- C Permission from the police
- D Permission from the compressor hire company
- E Working lights and a number plate on the compressor to match the towing vehicle

20.20

Who is responsible for the security of the load on a vehicle?

- A The driver's supervisor
- B Police
- C The driver
- D The driver's company

20.21

What is the correct way to dismount from the cab or the driving seat of plant and vehicles?

- A Use three points of contact facing forwards (away from the vehicle)
- B Jump down facing forwards (away from the vehicle)
- C Use three points of contact facing backwards (towards the vehicle)
- D Jump down well clear of the vehicle

20.22

Why is it necessary to wear hi-vis clothing when working on roads?

- A So road users and plant operators can see you
- B So that your supervisor can see you
- C Because you were told to
- D Because it will keep you warm

20.23

When working on a dual carriageway with a 60 mph speed limit, unless in the working space, what is the minimum standard of hi-vis clothing that must be worn?

- A Reflective waistcoat
- B Reflective long-sleeved jacket
- C Reflective sash
- D None

20.24

When working on the highway, hi-vis personal protective equipment (PPE) clothing gives you maximum protection when it is:

- A tied around your waist
- B worn open to flap in the wind
- C worn over other garments and fastened at the front
- D soiled and very dirty

F
20

20.25

When carrying out kerbing works, which method should be used for getting kerbs off the vehicle?

A Lift them off manually using the correct technique

B Push them off the back

C Use mechanical means, such as a JCB fitted with a grab

D Ask your workmate to give you a hand

20.26

What is the purpose of an on-site risk assessment?

A To ensure there is no risk of traffic build-up due to the works in progress

B To identify hazards and risks in order to ensure a safe system of work

C To ensure that the work can be carried out in reasonable safety

D To protect the employer from prosecution

20.27

In which TWO places would you find information on the distances for setting out highways signs in advance of the works under different road conditions?

A In the Traffic Signs Manual (Chapter 8)

B In the 'Pink Book'

C On the back of the sign

D In the specification for highway works

E In the Code of Practice ('Red Book')

20.28

You are the driver of a vehicle and lose sight of the signaller while reversing your vehicle. What do you do?

A Continue reversing slowly

B Continue reversing provided your vehicle is equipped with a klaxon and flashing lights

C Stop and locate the signaller

D Find someone else to watch you reverse safely

20.29

When materials have to be kept on site overnight, what should you do?

A Don't stack them above 2 m high

B Stack the materials safely and in a secure area

C Put pins and bunting around them

D Only stack them on the grass verge

20.30

What TWO actions should be taken to prevent the inhalation of harmful silica dust while mechanically cutting kerbs, slabs and blocks?

A Pump water onto the cutting disc to suppress the dust (wet cut)

B Press down on the disc cutter to finish the job faster

C No action is necessary as wind clears the dust

D Wear a correctly fitting mask with the correct dust protection factor (FFP3)

E Use a diamond blade cutting disc

20.31

When providing portable traffic signals on roads used by cyclists and horse riders, what action should you take?

A Locate the signals at bends in the road

B Allow more time for slow-moving traffic by increasing the 'all red' phase of the signals

C Operate the signals manually

D Use 'stop/go' boards only

20.32

Why should temporary highways signing be removed when works are complete?

A To get traffic flowing

B It is a legal requirement

C To allow the road to be opened fully

D To reuse signs on new jobs

20.33

When should installed highways signs and guarding equipment be inspected?

A After it has been used

B Once a week

C Before being used

D Regularly and at least once every day

20.34

What TWO site conditions must exist so that the minimum traffic management can be used?

A Traffic is heavy

B Visibility is good

C There are double yellow lines

D There is a footpath

E It is a period of low risk

20.35

What traffic management is required when carrying out a maintenance job on a motorway?

A The same as would be required on a single carriageway

B A flashing beacon and a 'keep left/right' sign

C A scheme installed by a registered traffic management contractor

D Ten one-metre high cones and a one-metre high 'men working' sign

F
20

20.36

What is the minimum traffic management required when carrying out a short-term minor maintenance job in a quiet, low-speed side road?

(A) A flashing amber beacon and a 'keep left/right' arrow

(B) The same as required for a road excavation

(C) Five cones and a blue arrow

(D) Temporary traffic lights

20.37

Some work activities move along the carriageway, such as sweeping, verge mowing and road lining. What is the maximum distance between the 'roadworks ahead' signs?

(A) Two miles

(B) One mile

(C) Half a mile

(D) Quarter of a mile

20.38

What action is required when a highways vehicle fitted with a direction arrow is travelling from site to site?

(A) Point the direction arrow up

(B) Travel slowly from site to site

(C) Point the direction arrow down

(D) Cover or remove the direction arrow

20.39

Signs placed on footways must be located so that they:

(A) block the footway

(B) can be read by site personnel

(C) do not create a hazard for pedestrians

(D) can be easily removed

20.40

What should you do if drivers approaching roadworks cannot see the advance signs clearly because of poor visibility or obstructions caused by road features?

(A) Place additional signs in advance of the works

(B) Extend the safety zones

(C) Extend the sideways clearance

(D) Lengthen the lead-in taper

20.41

How should a portable traffic-light cable, that crosses a road, be protected?

(A) It should be secured firmly to the road surface

(B) A cable crossing protector must be used with 'ramp' warning signs

(C) It can be unprotected if less than 10 mm diameter

(D) It should be placed in a slot cut into the road surface

F
20

20.42

What action is required where passing traffic may block the view of highways signs?

- **A** Signs must be larger
- **B** Signs must be duplicated on both sides of the road
- **C** Signs must be placed higher
- **D** Additional signs must be placed in advance of the works

20.43

Highways signs, lights and guarding equipment must be properly secured:

- **A** with sacks containing fine, granular material set at a low level
- **B** by roping them to concrete blocks or kerb stones
- **C** to prevent them being stolen
- **D** by iron weights suspended from the frame by chains or other strong material

20.44

Which is NOT an approved means of controlling traffic at roadworks?

- **A** Priority signs
- **B** Police supervision
- **C** Hand signals by operatives
- **D** A give-and-take system

20.45

What action is required if a vehicle detector on temporary traffic lights becomes defective?

- **A** Control traffic at the defective end using hand signals
- **B** Operate on 'all red' and call the service engineer
- **C** Operate on 'fixed time' or 'manual' and call the service engineer
- **D** Switch the lights off until the supervisor arrives on site

20.46

Portable traffic signals are assembled and placed:

- **A** as speedily as possible
- **B** in an organised manner to a specified sequence
- **C** during the night
- **D** as work starts each morning

20.47

What action is required where it is not possible to maintain the correct safety zone?

- **A** Barrier off the working space
- **B** Place additional advance signing
- **C** Use extra cones on the lead-in taper
- **D** Stop work and consult your supervisor

F
20

Highway works

20.48

In which of the following circumstances can someone enter the safety zone?

A To store unused plant

B To maintain cones and signs

C To park site vehicles

D To store materials

20.49

On a dual carriageway, a vehicle driven by a member of the public enters the coned-off area. What action do you take?

A Remove a cone and direct the driver back on to the live carriageway

B Ignore them

C Shout and wave them off site

D Assist them to leave the site safely via the nearest designated exit

20.50

If you are working after dark, is mobile plant exempt from the requirement to show lights?

A Yes, on all occasions

B Yes, if authorised by the supervisor

C Only if they are not fitted to the machine as standard

D Not in any circumstances

20.51

What is the purpose of the 'safety zone'?

A To indicate the works area

B To protect you from the traffic and the traffic from you

C To allow extra working space in an emergency

D To give a safe route around the working area

20.52

What should be used to protect the public from a shallow excavation in a public footway?

A Pins and bunting

B Nothing

C Cones

D Barriers with tapping rails

20.53

In which of the following circumstances would it NOT be safe to use a cherry picker for working at height?

A When a roll-over cage is not fitted

B When the ground is uneven and sloping

C When weather protection is not fitted

D When the operator is clipped to an anchorage point in the basket

F
20

21.1

If you need to store materials on a roof what THREE things must you do?

[A] Check the load bearing capability of the roof to avoid damage to the structure

[B] Stack materials no more than 1.2 metres above the guard-rail height

[C] Ensure there is safe access and clear working areas around the materials for everyone working on the roof

[D] Use a gin wheel and rope tied to a temporary tripod at the roof edge for raising and lowering the materials

[E] Store the materials in a way that prevents them from falling off or being blown off the roof

21.2

If a safety lanyard has damaged stitching, you should:

[A] use the lanyard if the damaged stitching is less than two inches long

[B] get a replacement lanyard

[C] not use the damaged lanyard and work without one

[D] use the lanyard if the damaged stitching is less than six inches long

21.3

What is the MAIN danger of leaving someone who has fallen suspended in a harness for too long?

[A] The anchorage point may fail

[B] They may try to climb back up the structure and fall again

[C] They may suffer severe trauma or even death

[D] It is a distraction for other workers

21.4

If using inflatable airbags as a means of fall arrest, the inflation pump must:

[A] be electrically powered

[B] be switched off from time to time to avoid over-inflation

[C] run all the time while work is carried out at height

[D] be switched off when the airbags are full

21.5

Why is it dangerous to use inflatable airbags that are too big for the area to be protected?

[A] They will exert a sideways pressure on anything that is containing them

[B] The pressure in the bags will cause them to burst

[C] The inflation pump will become overloaded

[D] They will not fully inflate

F
21

21.6

When is it most appropriate to use a safety harness and lanyard for working at height?

A Only when the roof has a steep pitch

B Only when crossing a flat roof with clear roof lights

C Only when all other options for fall prevention have been ruled out

D Only when materials are stored at height

21.7

When trying to clip your lanyard to an anchor point you find the locking device does not work. What should you do?

A Carry on working and report it later

B Tie the lanyard in a knot round the anchor

C Stop work and report it to your supervisor

D Carry on working without it

21.8

What is the main reason for using a safety net or other soft-landing system rather than a personal fall-arrest system?

A Soft-landing systems are cheaper to use and do not need inspecting

B It is always easy to rescue workers who fall into a soft-landing system

C Specialist training is NOT required to install soft-landing systems

D Soft-landing systems are collective fall arrest measures

21.9

Edge protection is designed to:

A make access to the roof easier

B secure tools and materials close to the edge

C stop rainwater running off the roof onto workers below

D prevent people and materials falling

21.10

On a working platform, the maximum permitted gap between the guard-rails is:

A 350 mm

B 470 mm

C 490 mm

D 510 mm

21.11

When should guard-rails be fitted to a working platform?

A If it is possible to fall 2 m

B At any height if a fall could result in an injury

C If it is possible to fall 3 m

D Only if materials are being stored on the working platform

F
21

Answers: 21.6 = C   21.7 = C   21.8 = D   21.9 = D   21.10 = B   21.11 = B

21.12

The Beaufort Scale is important when working at height externally because it measures:

A air temperature

B the load-bearing capacity of a flat roof

C wind speed

D the load-bearing capacity of a scaffold

21.13

Before starting work at height, the weather forecast says the wind will increase to 'Force 7'. What does this mean?

A A moderate breeze that can raise light objects, such as dust and leaves

B A near gale that will make it difficult to move about and handle materials

C A gentle breeze that you can feel on your face

D Hurricane winds that will uproot trees and cause structural damage

21.14

You have to lean over an exposed edge while working at height. How should you wear your safety helmet?

A Titled back on your head so that it doesn't fall off

B Take your helmet off while leaning over then put it on again afterwards

C Wear the helmet as usual but use the chinstrap

D Wear the helmet back to front whilst leaning over

21.15

Before climbing a ladder you notice that it has a rung missing near the top. What should you do?

A Do not use the ladder and immediately report the defect

B Use the ladder but take care when stepping over the position of the missing rung

C Turn the ladder over so that the missing rung is near the bottom and use it

D See if you can find a piece of wood to replace the rung

21.16

How far should a ladder extend above the stepping-off point if there is no alternative, firm handhold?

A Three rungs

B Two rungs

C One metre

D Half a metre

F
21

21.17

When using portable or pole ladders for access, what is the maximum vertical distance between landings?

- A 5 m
- B There is no maximum
- C 9 m
- D 30 m

21.18

A 'Class 3' ladder is:

- A for domestic use only and must not be used on site
- B of industrial quality and can be used safely
- C a ladder that has been made to a European Standard
- D made of insulating material and can be used near to overhead cables

21.19

You need to use a ladder to access a roof. The only place to rest the ladder is on a run of plastic gutter. What TWO things should you consider doing?

- A Rest the ladder on a gutter support bracket
- B Rest the ladder against the gutter, climb it and quickly tie it off
- C Find another way to access the roof
- D Use a proprietary 'stand-off' device that allows the ladder to rest against the wall
- E Position the ladder at a shallow angle so that it rests below the gutter

21.20

Scaffold towers should only be erected or dismantled by:

- A someone who has the instruction book
- B someone who is trained, competent and authorised
- C advanced scaffolders
- D someone who has worked on them before

21.21

What is the recommended maximum height for a free-standing mobile tower when used indoors?

- A There is no restriction
- B Three lifts
- C In accordance with the manufacturer's recommendations
- D Three times the longest base dimension

21.22

After gaining access to the platform of a mobile tower, the first thing you should do is:

- A check that the tower's brakes are locked on
- B check that the tower has been correctly assembled
- C close the access hatch to stop people or equipment from falling
- D check that the tower does not rock or wobble

F
21

21.23

Before a mobile tower is moved, you must first:

A clear the platform of people and equipment

B get a permit to work

C get approval from the principal contractor

D make arrangements with the forklift truck driver

21.24

An outdoor tower scaffold has stood overnight in high winds and heavy rain. What should you consider before the scaffold is used?

A That the brakes still work

B Tying the scaffold to the adjacent structure

C That the scaffold is inspected by a competent person

D That the platform hatch still works correctly

21.25

If working from a cherry picker you should attach your safety lanyard to a:

A strong part of the structure you are working on

B secure anchorage point inside the platform

C secure point on the boom of the machine

D scaffold guard-rail

21.26

You are working at height from a cherry picker when the weather becomes very windy. Your first consideration should be to:

A tie all lightweight objects to the handrails of the basket

B clip your lanyard to the structure that you are working on

C tie the cherry picker basket to the structure you are working on

D decide whether the machine will remain stable

21.27

You are on a cherry picker but it does not quite reach where you need to work. What should you do?

A Use a stepladder balanced on the machine platform

B Extend the machine fully and stand on the guard-rails

C Abandon the machine and use a long extending ladder

D Do not carry out the job until you have an alternative means of access

F
21

21.28

If you are working at height and operating a mobile elevating work platform (MEWP), when is it acceptable for someone to use the ground-level controls?

- A If the person on the ground is trained and you are not
- B In an emergency only
- C If you need to jump off the MEWP to gain access to the work
- D If you need both hands free to carry out the job

21.29

When is it acceptable to jump off a mobile elevating work platform (MEWP) on to a high level work platform?

- A If the work platform is fitted with edge protection
- B If the machine operator stays in the basket
- C Not under any circumstances
- D If the machine is being operated from the ground-level controls

21.30

How will you know the maximum weight or number of people that can be lifted safely on a mobile elevating work platform (MEWP)?

- A The weight limit is reached when the platform is full
- B It will say on the Health and Safety Law poster
- C You will be told during site induction
- D From an information plate fixed to the machine

21.31

When is it safe to use a scissor lift on soft ground?

- A When the ground is dry
- B When the machine can stand on scaffold planks laid over the soft ground
- C When stabilisers or outriggers can be deployed onto solid ground
- D Never

21.32

You need to cross a roof. How do you establish if it is fragile?

- A Tread gently and listen for cracking
- B Look at the risk assessment or method statement
- C Look at the roof surface and make your own assessment
- D It does not matter if you walk along a line of bolts

F
21

Answers: 21.28 = B   21.29 = C   21.30 = D   21.31 = C   21.32 = B

21.33

After gaining access to a roof, you notice some overhead cables within reach. What should you do?

A. Keep away from them while you work but remember they are there

B. Confirm that it is safe for you to be on the roof

C. Make sure that you are using a wooden ladder

D. Hang coloured bunting from them to remind you they are there

21.34

What should be included in a safety method statement for working at height? Give THREE answers.

A. The cost of the job and time it will take

B. The sequence of operations and the equipment to be used

C. How much insurance cover will be required

D. How falls are to be prevented

E. Who will supervise the job on site

21.35

When working above public areas, your first consideration should be to:

A. minimise the number of people below at any one time

B. prevent complaints from the public

C. let the public know what you are doing

D. prevent anything falling on to people below

21.36

When covering roof lights, what TWO requirements should the covers meet?

A. They are made from the same material as the roof covering

B. They are made from clear material to allow the light through

C. They are strong enough to take the weight of any load placed on them

D. They are waterproof and windproof

E. They are fixed in position to stop them being dislodged

21.37

Which of these must happen before any roof work starts?

A. A risk assessment must be carried out following the hierarchy of risk

B. The operatives working on the roof must be trained in the use of safety harnesses

C. Permits to work must be issued to those allowed to work on the roof

D. A weather forecast must be obtained

F
21

Answers: 21.33 = B   21.34 = B, D, E   21.35 = D   21.36 = C, E   21.37 = A

21.38

When working at height, which of these is the safest way to transfer waste materials to ground level?

A) Though a waste chute directly into a skip

B) Ask someone below to keep the area clear of people then throw the waste down

C) Erect barriers around the area where the waste will land

D) Bag or bundle up the waste before throwing it down

21.39

You need to inspect pipework at high level above an asbestos roof. You should:

A) use an extension ladder and crawler board to get to the pipework

B) use a ladder to get onto the roof and walk the bolt line on the roof sheets

C) report the pipework as unsafe

D) hire in suitable mobile access equipment

21.40

You have been asked to erect specialist access frames using anchor bolts. Before you start work what should you NOT do?

A) Check the access frames are sound

B) Assume that the access system is safe to use

C) Test the anchor bolts

D) Ensure that your assistant has their harness on

21.41

You are working above a safety net. You notice the net is damaged. What should you do?

A) Work somewhere away from the damaged area of net

B) Stop work and report it

C) Tie the damaged edges together using the net test cords

D) See if you can get hold of a harness and lanyard

21.42

What is the MAIN reason for not allowing debris to gather in safety nets?

A) It will overload the net

B) It looks untidy from below

C) It could injure someone who falls into the net

D) Small pieces of debris may fall through the net

21.43

You are working at height, but the securing cord for a safety net is in your way. What should you do?

A) Untie the cord, carry out your work and tie it up again

B) Untie the cord, but ask the net riggers to re-tie it when you have finished

C) Tell the net riggers that you are going to untie the cord

D) Leave the cord alone and report the problem

F
21

21.44

Ideally, a safety net should be rigged:

A immediately below where you are working

B 2 m below where you are working

C 6 m below where you are working

D at any height below the working position

21.45

Who should install safety nets?

A A scaffolder

B Someone who has had experience of working with them before

C A trained, competent and authorised person

D A steel or cladding erector

21.46

When can someone who is not a scaffolder remove parts of a scaffold?

A If the scaffold is not more than two lifts in height

B As long as a scaffolder refits the parts after the work has finished

C Never, only competent scaffolders can remove the parts

D Only if they think the parts won't weaken the scaffold

21.47

During your work, you find that a scaffold tie is in your way. What should you do?

A Ask a scaffolder to remove it

B Remove it yourself and then replace it

C Remove it yourself but get a scaffolder to replace it

D Report the problem to your supervisor

21.48

Which type of scaffold tie can be removed by someone who is not a scaffolder?

A A box tie

B A ring tie

C A reveal tie

D None

F
21

22.1

Who is permitted to undertake the safe release of trapped passengers?

- A The site foreman
- B Only a trained and authorised person
- C Anyone
- D Only the emergency services

22.2

You need to connect a car light supply to a 240 volt supply (240 volt fused spur) supplied by others. Do you:

- A connect with the power on
- B switch off the spur and connect
- C switch off the spur and remove the fuse and connect
- D isolate and lock off the incoming supply and connect

22.3

A switch needs to be changed in the pit and the isolator is in the machine room 12 floors above. Do you:

- A isolate the power and lock and tag the isolator
- B risk assess the situation and, because it's control voltage, do it with the power on
- C use insulated tools
- D stand on a rubber mat

22.4

What is the main cause of injury and absence to workers within the lift and escalator industry?

- A Falls
- B Electrocution
- C Contact with moving parts
- D Manual handling

22.5

If a counterweight screen is not fitted or has been removed, what should you do before starting work?

- A Carry out a further risk assessment to establish a safe system of work
- B Nothing – just get on with the job as normal
- C Give a toolbox talk on guarding
- D Issue and wear appropriate personal protective equipment (PPE)

22.6

Which of the following types of fire extinguisher should NOT be used if a fire occurs in a lift or escalator controller?

- A Halon
- B Water
- C Dry powder
- D Carbon dioxide

22.7

What should you do if the lifting tackle you are about to use is defective?

A) Only use it for half its safe working load

B) Only use it for small lifts under 1 tonne

C) Do not use it and inform your supervisor

D) Try to fix it

22.8

If landing doors are not fitted to a lift on a construction site, what is the minimum height of the barrier that must be fitted instead?

A) 650 mm

B) 740 mm

C) 810 mm

D) 950 mm

22.9

A set of chain blocks has been delivered to site with a certificate stating they were inspected by a competent person a month before. The hook is obviously damaged. What action do you take?

A) Use the blocks as the certificate is in date

B) Do not use the blocks and inform your supervisor

C) Use the blocks at half the safe working load

D) Use the blocks until replacement equipment arrives

22.10

Rings, bracelets, wrist watches, necklaces and similar items must NOT be worn:

A) when working near or on electrical or moving equipment

B) when working on site generally

C) when driving a company vehicle

D) after leaving home for work

22.11

What is the correct method of disposal for used or contaminated oil?

A) Decant it into a sealed container and place in a skip

B) Through a registered waste process

C) Dilute it with water and pour down a sink

D) Pour it down a roadside drain

22.12

A large, heavy, balance weight frame is delivered to site on a lorry with no crane. There is no lifting equipment available on site. Do you:

A) unload it manually

B) arrange for it to be re-delivered on a suitable lorry

C) slide it down planks

D) tip the load off

F
22

Lifts and escalators

22.13

A lifting beam at the top of the lift shaft is marked with a safe working load of 800 kg but the brickwork around the beam is cracked and appears to be loose. Do you:

- A use the beam as normal
- B only lift loads not exceeding 400 kg
- C not use the beam and speak to your supervisor
- D de-rate the beam by 75%

22.14

To prevent injury from an overspeed governor, what is fitted?

- A A rope
- B A restrictor
- C A guard
- D A switch

22.15

If the escalator or passenger conveyor has an external machine room, access doors should be:

- A capable of being locked and be marked with the appropriate safety sign
- B smoke proof in case of a fire
- C unlocked at all times in case of an emergency
- D capable of being locked on the inside only and be marked with the appropriate safety sign

22.16

What is the frequency of the statutory period of inspection for lifting equipment used to support people?

- A At least monthly
- B At least every six months
- C At least every 12 months
- D Once every two years

22.17

What checks do you need to carry out before using lifting equipment?

- A A drop check
- B It is free from defects and has a current inspection certificate
- C Ensure the chains are knotted to the correct length
- D The date of manufacture of the lifting tackle

22.18

Following the initial inspection, how often should a scaffold in a lift shaft be inspected by a competent person?

- A At least every day
- B At least every seven days
- C At least every 14 days
- D There is no set period between inspections

F
22

22.19

While installing a new rope, you notice a damaged section where something heavy has fallen onto the coil. Do you:

A fit the rope anyway

B cut out the damaged section

C reject the rope

D add an extra termination

22.20

Before entering the pit of an operating lift you must first:

A fit pit props

B verify the pit stop switch

C switch the lift off

D position the access ladder

22.21

Who should fit a padlock and tag to an electrical lock-out guard?

A Anyone working on the unit

B The engineer who fitted the lock-out device

C The senior engineer

D Anyone

22.22

You arrive on site and find the lift mains isolator is switched off. What should you do?

A Switch it on and get on with your work

B Switch it on and check the safety circuits to see if there is a fault

C Contact the person in control of the premises to find out if they had switched the lift off or if anyone else is working on site

D Shout down the shaft and if no-one responds switch it on and get on with your work

22.23

It is essential that an authorised person working alone does which TWO of the following?

A Before starting work, registers their presence with the site representative

B Ensures their timesheet is accurate and countersigned

C Establishes suitable arrangements to ensure the monitoring of their wellbeing

D Notifies the site manager of the details of the work

E Ensures that the lift pit is free from water and debris

F
22

Lifts and escalators

22.24

When using authorised lifting tackle marked with its safe working load, which statement is true?

- A) Never exceed the safe working load
- B) The safe working load is only guidance
- C) Halve the safe working load if the equipment is damaged
- D) Double the safe working load if people are to be lifted

22.25

What should be fitted to the main sheave and diverter to prevent injury from rotating equipment?

- A) Movement sensors
- B) Guards
- C) Clutching assembly
- D) Safety notices

22.26

When working on electrical lift-control equipment what are the appropriate tools and equipment?

- A) Insulated tools and an insulating mat
- B) Non-insulated tools
- C) Any tools and an insulating mat
- D) No tools are allowed near electrical equipment

22.27

When installing a partially enclosed or observation lift, what safe system of work can you use to prevent injury to people below?

- A) Put up a sign
- B) Do not use heavy tools
- C) Secure tools to prevent them falling off
- D) Only carry out essential work using minimum tools

22.28

What needs to be checked before any hot work takes place in the lift installation?

- A) The weight and size of the welding equipment
- B) How long the task will take
- C) If a 'hot work' permit or permit to work is required
- D) If the local fire services require notifying

22.29

If the trapdoor or hatch has to be left open while you work in the machine room, what must you ensure?

- A) That a sign is posted to warn others
- B) That the distance from the trapdoor/hatch to the floor below does not exceed 2 m
- C) That there is sufficient light available for the work
- D) That a suitable barrier is put in place

F
22

22.30

To prevent unauthorised access to unoccupied machine equipment space, what must you ensure?

- [A] That the access door is locked
- [B] That a sign is posted
- [C] That the power supply is isolated
- [D] That a person is posted to prevent access

22.31

What should be applied to the main isolator of a traction lift to prevent accidental starting?

- [A] A warning notice
- [B] Lock-out device
- [C] RCD (residual current device)
- [D] Lower-rated fuses

22.32

Who is responsible for the keys when a padlock has been applied to a lock-out device?

- [A] The individual applying the lock
- [B] Supervisor
- [C] Manager
- [D] The person nearest the lock-out device

22.33

If work is to be done on electrical lift equipment, and the main isolator does not have a lock-out device, what is an alternative method of isolating the supply?

- [A] Detach the main power supply
- [B] Place insulation material in the contactors
- [C] There is no alternative to a lock-out device for isolating the supply
- [D] Withdraw and retain the fuse and fix notices warning that the machinery is being worked on

22.34

The main contractor wants to use the unfinished lift to move some equipment to an upper floor. What should you do?

- [A] Help to ensure the load is correctly positioned
- [B] Refer them to your supervisor
- [C] Ask for the weight of the equipment
- [D] Allow them to use the lift but take no responsibility

22.35

Before gaining access into the escalator or passenger conveyor, it is essential that:

- [A] the mains switch is locked and tagged out
- [B] the mains switch is in the 'On' position
- [C] all steps are removed
- [D] the drive mechanism is lubricated

F 22

22.36

What is secured at the entry/exit points of an escalator/passenger conveyor to prevent people falling into the machine or machine space?

A) Safety barriers

B) Safety notices

C) Escalator machine equipment guards

D) Machine tank covers

22.37

What must you do before moving the steps or pallet band of an escalator or passenger conveyor?

A) Check there are no sharp edges on the steps

B) Check there is a clear route of escape

C) Check that no unauthorised people are on the equipment

D) Check that a fire extinguisher is available

22.38

A gap of what size or more, between the edge of the work platform and the hoist way wall, is regarded as a fall hazard?

A) 250 mm

B) 300 mm

C) 330 mm

D) 450 mm

22.39

When is it acceptable to work on the top of a car without a top-of-car control station?

A) When the unit has been locked and tagged out

B) When there are two engineers

C) When there is no other way

D) When single-person working

22.40

When working on an energised car, which statement is true?

A) Always attach your lanyard to the car top while standing on the landing

B) Make sure you step onto the landing with your lanyard still attached to the car

C) Always check the lanyard is unclipped before getting off the car top

D) Ensure that your lanyard is clipped to a guide bracket or similar anchorage on the shaft

F
22

22.41

What is the last thing you should do before alighting from a car top through open landing doors when the car-top control is within 1 m of the landing threshold?

A. Set the car-top control to 'Test'

B. Ensure that the car-top stop button is set to 'Stop' and the car-top control remains set to 'Test'

C. Turn off the shaft lights and switch the car-top control to 'Normal' operation

D. Press the stop button and switch the car-top control to 'Normal' operation

22.42

If the landing doors are to be open while work goes on in the lift pit, what precaution must you take?

A. Erect a suitable barrier and secure it in front of the landing doors

B. Post a notice on the wall next to the doors

C. Do the job when there are few people about

D. Ask someone to guard the open doors while you work

22.43

Before accessing the car top of an operating lift, what is the first thing to do after opening the landing doors?

A. Make sure the lift has stopped

B. Press the car-top stop button

C. Chock the landing doors open

D. Put the car-top control in the 'Test' position

22.44

When you gain access to a car top, you should test that the car-top stop switch operates correctly by:

A. trying to move the car in the up direction

B. trying to move the car in the down direction

C. measuring with a multimeter

D. flicking the switch on and off rapidly

22.45

When working in the pit, the lift should be positioned towards the top of the shaft unless:

A. the hydraulic fluid level is low

B. work needs to be done on the underside of the lift

C. you are testing the buffers

D. the power supply is cut

F
22

22.46

What is required on each landing of a new lift shaft before entrances and doors are fitted?

 A warning notice

B A substantial secure barrier to prevent falls

C Orange plastic netting across the opening

D Lighting

22.47

When handling stainless steel car panels, which of the following items of personal protective equipment (PPE) should you wear in addition to safety footwear?

A Suitable safety gloves, such as rigger type gloves

B Hand barrier cream

C Latex gloves

 D Hearing protection

22.48

At what stage in the installation of a lift should guarding be fitted to the lift machine?

 A At the end of the job

B During commissioning

C When handing over to the client

D Before the machine can be operated

22.49

Which statement is INCORRECT?

A A stop switch must be within 1.5 m of the front of the car

B The car top should be clean and free from grease and oil spills

C You should secure your tools out of your standing area when working on top of the car

D Before trying to gain access to the hoist way, you should decide whether the work will need the power supply to be live

22.50

If it is unavoidable that one person is to work above another in a lift shaft what can you do to reduce the risk of the person below being harmed by falling objects?

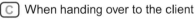 A Minimise the tools and equipment in use above

B Have tools on lanyards

C Use safety helmets

D All of these answers

F
22

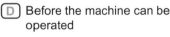

23.1

When working underground where should your self-rescue set be?

- A On your person
- B Immediately available
- C At the nearest fire point
- D Within two minutes' walking distance

23.2

The duration of a self-rescue set would be significantly reduced if the wearer uses it while:

- A running
- B waiting for rescue
- C standing still
- D smoking

23.3

BS 6164 states a self-rescue set used for escape should have a supply of air lasting at least:

- A 5 minutes
- B 10 minutes
- C 20 minutes
- D 30 minutes

23.4

In the event of an emergency who will use the information displayed on a tally board?

- A Health and Safety Executive (HSE)
- B DSS
- C Rescue services
- D Environmental Health

23.5

In the event of discovering a fire whilst in a tunnel, you should:

- A leave the tunnel immediately
- B leave the tunnel immediately and then raise the alarm
- C contact your supervisor and report the fire's location
- D raise the alarm and leave the tunnel, if safe to do so

23.6

Entry and exit of all personnel with regard to a confined space should be controlled by:

- A CCTV
- B light signals
- C a tally system
- D verbal communication with the top man

23.7

Which of the following is NOT permitted underground?

- A The use of diesel
- B Smoking or smoking materials
- C Welding equipment
- D Gas bottles

F
23

Answers: 23.1 = B   23.2 = A   23.3 = C   23.4 = C   23.5 = D   23.6 = C   23.7 = B   **157**

Tunnelling

23.8

What action should you take if the ventilation system alarm activates?

- A Carry on working until advised otherwise
- B Stop work and take a break
- C Evacuate the workplace
- D Ask your workmates

23.9

The atmosphere is oxygen deficient when the level of oxygen falls below:

- A 18%
- B 19%
- C 20%
- D 21%

23.10

Petrol-operated equipment:

- A is not allowed underground
- B can only be used for underground lighting
- C is used for all power sources
- D can be used with an exhaust filter

23.11

Intrinsically safe equipment should be used in the underground environment where there is a risk of encountering:

- A water
- B potentially explosive atmosphere
- C air
- D poisonous gas

23.12

All diesel engine plant used underground should be fitted with a:

- A toilet
- B fire detection system
- C fixed extinguishing system
- D portable extinguishing system

23.13

An oxygen-deficient atmosphere will leave you:

- A feeling happy
- B breathless
- C acting drunk
- D feeling excited

23.14

Underground petrol-operated equipment is:

- A used intermittently to prevent a build-up of fumes
- B the only equipment to be used
- C not to be used at all
- D used only if fitted with exhaust filters

F 23

Answers: 23.8 = C   23.9 = B   23.10 = A   23.11 = B   23.12 = C   23.13 = B
23.14 = C

23.15

Methane gas does not have an odour. In what TWO ways can it be dangerous?

- A It causes skin irritation
- B It can cause temporary blindness
- C It is explosive
- D It is toxic
- E It reduces oxygen in the atmosphere

23.16

Which of the following is one way to tell that there may be hydrogen sulphide in the air?

- A There is a smell of rotten eggs
- B There is a yellow dust
- C A mist forms
- D You feel excited

23.17

Which of the following is a reliable way to detect carbon monoxide and methane gas?

- A They both have a distinctive bad egg smell
- B They are both harmless
- C With a calibrated gas detector
- D They both form a mist

23.18

How could hydrogen sulphide affect those working in a tunnel?

- A It can leave a yellow dust which can irritate the skin
- B It can make the atmosphere explosive
- C It can cause a mist making it difficult to see
- D It can cause death through stopping breathing

23.19

Carbon monoxide is dangerous because it:

- A is soluble in water
- B is a skin irritant
- C is an explosive
- D stops the intake of oxygen

23.20

What should indicate that the ventilation system has stopped working?

- A A notice at the daily briefing
- B You are informed as part of your induction
- C An audible alarm sounds
- D Your supervisor would inform you

F
23

23.21

Nitrogen oxide (NO) gas can be present in tunnels. Which one of these plant items causes most nitrogen oxide to be generated/made?

- [A] Tunnel-boring machines (TBMs)
- [B] Electro-hydraulic spray pumps
- [C] Rail-mounted plant
- [D] Diesel-powered equipment

23.22

Exposure to nitrogen oxide (NO) gas can cause breathing problems. Which of the following is the PREFERRED control measure?

- [A] Provide air monitoring
- [B] Have a portable breathing set nearby
- [C] Avoid exposure
- [D] Provide ventilation systems

23.23

The power supply for communication equipment should be:

- [A] linked to the main tunnel power supply
- [B] independent of the main tunnel power supply
- [C] linked to the 33 kVA power supply
- [D] battery-powered

23.24

What is a common method used for communication between the tunnelling face and the surface?

- [A] Email
- [B] A two-way radio or tannoy system
- [C] Mobile phones
- [D] Text messages

23.25

The distance between emergency lighting in a tunnel should NOT exceed:

- [A] 25 m
- [B] 50 m
- [C] 75 m
- [D] 100 m

23.26

Mains-powered, electrically operated, portable hand tools must be:

- [A] 42 volts
- [B] 110 volts
- [C] 240 volts
- [D] 415 volts

23.27

What is the colour of a 415 volt plug?

- [A] Black
- [B] Blue
- [C] Yellow
- [D] Red

F
23

Answers: 23.21 = D   23.22 = C   23.23 = B   23.24 = B   23.25 = B   23.26 = B
23.27 = D

23.28

When working in a tunnel:

A smoking can take place in all underground tunnels

B smoking materials can be taken underground

C smoking materials are not permitted underground

D only cigarette smoking is permitted

23.29

Why is smoking not allowed underground?

A Smoking is damaging to health

B Smoking is a fire hazard

C Non-smoking workers have complained

D You can't smoke while wearing personal protective equipment (PPE)

23.30

When charging lead acid batteries they produce an explosive gas, smoking or other naked flames are NOT permitted within what distance of a battery-charging area?

A 5 m

B 10 m

C 15 m

D 20 m

23.31

Hot work activities are not permitted within what distance of a diesel-fuelling point?

A 5 m

B 10 m

C 15 m

D 20 m

23.32

As a minimum, how long must a fire watch be maintained after hot works have been completed?

A 15 minutes

B 30 minutes

C 45 minutes

D 60 minutes

23.33

The British Standard for tunnelling states that:

A smoking materials should be prohibited underground

B smoking is allowed anywhere below ground

C designated smoking areas should be set up below ground

D patches and gum should be provided for smokers working in tunnels

F
23

Answers: 23.28 = C   23.29 = B   23.30 = B   23.31 = B   23.32 = D   23.33 = A   **161**

23.34

In compressed air tunnelling the term DCI refers to:

 (A) dizzy, confused, induced state

(B) decompression incident

(C) decompression, combustion incident

(D) decompression illness

23.35

Which of the following characteristics are associated with hydrogen sulphide gas?

(A) A smell of rotten eggs

(B) It causes skin irritation

(C) It is odourless

(D) It causes a yellow haze

23.36

Which of the following statements is correct regarding hand-arm vibration syndrome?

 (A) It can be prevented if gloves are worn

 (B) It can be partially cured with medication

 (C) It can be corrected by surgery

(D) Once damage has happened it cannot be cured

23.37

Batching plants must be equipped with:

(A) spanners for required repairs

(B) sterile bandages

(C) a first aider

(D) an eyewash station

23.38

What main health risk is NOT usually associated with using sprayed concrete linings?

(A) Cement burns

(B) Arc eye

(C) Hand-arm vibration syndrome

(D) Inhalation of dust

23.39

What is the biggest risk to health associated with hydrogen sulphide gas?

(A) It can irritate the eyes and throat

(B) It causes skin irritation

(C) It causes hand-arm vibration syndrome

(D) It causes respiratory paralysis

23.40

Which of the following is NOT a hazard associated with hand mining?

(A) Vibrating hand tools

(B) Noise

(C) Falling mined material

(D) Sprayed concrete rebound

F
23

Answers: 23.34 = D   23.35 = A   23.36 = D   23.37 = D   23.38 = B   23.39 = D
23.40 = D

23.41

When should a hop up/refuge in the tunnel be used?

(A) When vehicles are passing

(B) For services such as cables

(C) When installing ventilation cassettes

(D) For storage of materials and equipment

23.42

Which of the following must be fitted as a conveyor system safety device?

(A) Emergency lighting

(B) Emergency pull cord or stop button

(C) Hazard lighting

(D) Seat

23.43

Secondary couplings must be fitted to unbraked rolling stock to reduce the risk of it:

(A) being derailed

(B) running away

(C) being overloaded

(D) stalling

23.44

A locomotive/vehicle is approaching you in the tunnel. When should you make your way to a hop up/safe refuge?

(A) Immediately

(B) Only when you can see it

(C) When you have identified the direction of travel

(D) Only when everyone else starts to move

23.45

Audible alarms are fitted to conveyors to warn those nearby that:

(A) it is about to stop

(B) it is about to break down

(C) it is about to start

(D) the belt needs replacing

23.46

A locomotive is entering the rear of the tunnel-boring machine. Which TWO electronic systems are recommended to assist in controlling its movements?

(A) Signal/traffic lights

(B) CCTV in the cab

(C) Siren

(D) Telephone

(E) Klaxon bell

F
23

23.47

Inclined conveyors are fitted with anti-roll back devices to prevent the belt running backwards due to which TWO potential failures?

A. Overloading

B. Power loss

C. Oil spillage

D. Overheating

E. Water leak

23.48

A common traffic light system used underground to control plant movement is:

A. red = stop
amber = out bye
green = in bye

B. red = stop
amber = in bye
green = out bye

C. red = in bye
amber = stop
green = out bye

D. red = out bye
amber = in bye
green = stop

23.49

The lights fitted on locomotives must be visible at a minimum distance of:

A. 50 m

B. 60 m

C. 70 m

D. 80 m

23.50

Alarms are fitted to a tunnel-boring machine's rams and erectors to warn workers when they:

A. are moving or about to move

B. are being maintained

C. have completed the tunnel bore

D. are stopping

23.51

A locomotive is travelling in the tunnel and you see a red light. This indicates that the locomotive:

A. has stopped

B. is travelling towards you

C. is travelling away from you

D. has broken down

23.52

What is the likely hazard from moving plant or locomotives in a tunnel?

A. Crush

B. Crash

C. Noise

D. Asphyxiation

23.53

Identify the LEAST effective method of controlling locomotive movements in the pit bottom.

A. Traffic lights

B. Radio

C. Shouting

D. Hand signals

F
23

Answers: 23.47 = A, B   23.48 = A   23.49 = B   23.50 = A   23.51 = C   23.52 = A
23.53 = C

23.54

In an emergency a locomotive must be able to stop within a distance not exceeding:

A 600 m

B 100 m

C 60 m

D 10 m

23.55

How often should safe-refuge (hop ups) be located along a tunnel?

A 50 m on straights;
25 m on curves

B 60 m on straights;
30 m on curves

C 70 m on straights;
25 m on curves

D 80 m on straights;
20 m on curves

23.56

The tunnel-boring machine operator has restricted vision during the building process. Which of the following is the most effective way of overcoming this?

A Alternative control point

B Mirrors at shoulder/crane level

C Use a signaller

D Get one of the gang to build

23.57

A locomotive is travelling in the tunnel and you see a white light. This means that the locomotive:

A has stopped

B is travelling towards you

C is travelling away from you

D has broken down

23.58

Identify a potential for accidents arising from pumped grouting operations.

A Injury through a hose bursting

B Potential hearing loss due to noise

C Injury through blowout at the injection point

D All of these answers

23.59

Blockages in grouting pipelines can cause the pipe to burst. What is the first action to be taken?

A Locate the blockage

B Split the line

C Clean out section by section

D Release the pressure in the pipeline

F
23

23.60

Why is it important to clean grouting pipelines after use to prevent injury to workers?

(A) It helps prevent blockages, which could cause the hose to burst

(B) It helps prevent the pipes from becoming weakened

(C) It keeps the dust level below the point that dust masks are needed

(D) It prevents the atmosphere from becoming explosive

23.61

Damaged flexible hoses can cause injury. One visible sign that a hose may be damaged is:

(A) swelling of the connections

(B) swelling of the hose

(C) difficulty in making a connection

(D) hardening of the rubber

23.62

Anti-whip devices must be fitted across:

(A) flexible hose connections

(B) rail connections

(C) steel pipe connections

(D) locomotive couplings

23.63

Anti-whip devices fitted to flexible hoses are designed to prevent:

(A) sudden parting of the connection under pressure

(B) a build up of dust

(C) hose ends flying about if they suddenly disconnect under pressure

(D) whole body vibration

23.64

To prevent injury during maintenance work on slurry lines a safe system of work should include which of the following?

(A) Isolation arrangements

(B) Report to the CDM co-ordinator

(C) Radar detection

(D) Unauthorised use

23.65

Which of the following statements is NOT true with regard to two hoses with different markings?

(A) They are different sizes

(B) They have different operating capacities

(C) They have different applications

(D) They can safely be joined together

F
23

23.66

Crash cages and side bars/or sliding doors should be fitted to all man-riding cars used in the tunnel. Select TWO reasons why they are needed.

 A To prevent derailment of the man-riding car

B To minimise injury in the event of a derailment

C To allow personnel to talk while being transported

D To stop personnel sitting close together

E To prevent personnel leaning out or falling out

23.67

Entry and exit to any tunnel under construction should be controlled by:

A CCTV

B a tally system

C verbal communication

D control is not necessary

23.68

For maintenance purposes water in the tunnel should NOT be allowed to accumulate at or above:

A axis level

B rail level

C walkway level

D crown level

23.69

Vertical ladders in shafts must have a landing interval every:

A 4 m

B 6 m

C 8 m

D 10 m

23.70

For tunnelling operations what is the minimum number of escape routes or methods that must be maintained from a working shaft?

A One

B Two

C Three

D Four

23.71

What term is used for the access and egress control system to tunnels?

A Visitor book

B Tally system

C Signing-in book

D Control is not needed

F
23

23.72

If a man-riding cage is used to transport workers into and out of a shaft, how many people can be transported at any one time?

A As defined by the Health and Safety Executive (HSE)

B As many as can fit into it

C As many as stated on the man-rider

D As many as stated by the supervisor

23.73

In straight sections of tunnel what is the recommended maximum distance between safe refuges?

A 25 m

B 50 m

C 75 m

D They are not required

23.74

On curved sections of tunnel what is the recommended maximum distance between safe refuges?

A Any distance

B 15 m

C 25 m

D 35 m

23.75

Compressed air tunnelling is used to:

A prevent water from leaving the tunnel

B avoid gas build up in the tunnel

C stop air escaping from the tunnel

D prevent/reduce water ingress during tunnel construction

23.76

Tunnel services should be correctly located to ensure:

A they cannot be seen

B they are in a safe position

C they are in an unsafe position

D it doesn't matter as any position works

23.77

Hydraulic jacking rams in a pipe-jacking shaft/pit should be fitted with:

A cushions to sit on

B audible/visual alarm to alert when moving

C the manufacturer's label

D a two-way radio

23.78

BS 6164 is the code of practice for safety in:

A construction

B tunnelling

C land management

D construction, design and management

F 23

Answers: 23.72 = C   23.73 = B   23.74 = C   23.75 = D   23.76 = B   23.77 = B
23.78 = B

23.79

Underground, a water spray curtain can help in slowing down the movement of:

A smoke

B unpleasant smells

C vermin

D waste skips

23.80

Segment erector equipment should have a fail to safety device to stop the operation in case of:

A malfunction, leakage or power failure

B electrocution

C anti-clockwise rotation

D overloading

23.81

A safe system of work for track maintenance must include which TWO of the following?

A A permit to work

B Red and green flags

C Adequate refuge space

D Crash deflectors

E A report for the CDM co-ordinator

23.82

A secure barrier must be maintained around any shaft to prevent falls. What is the minimum recommended height of the barrier?

A 0.9 m

B 1.2 m

C 1.5 m

D 1.8 m

23.83

What other controls should be used to warn locomotive drivers of personnel working in the tunnel?

A Permit system

B Trackside warning lights

C Lookouts and workers wearing hi-vis clothes

D All of these answers

23.84

Where a hazard cannot be eliminated the next control measure is for the risks to be:

A minimised

B disabled

C monitored

D accepted

F
23

23.85

Which of the following controls should be considered for shafts under construction?

A) Ventilation

B) Continuous gas monitoring

C) Shaft guarding

D) All of these answers

23.86

Stored tunnel segments should be secured by suitable means to prevent:

A) damage to the segments

B) additional loading on tunnel excavations

C) segments toppling over

D) all of these answers

23.87

Workers should not be allowed to work alone in tunnels because:

A) nobody will know whether they have had an accident and they may be unable to communicate

B) they may get lonely

C) the CDM co-ordinator has prevented it

D) they may lose track of time

23.88

Which of the following situations would require using a safety harness?

A) Working as the signaller at the pit top

B) Working as the belt-man on a TBM

C) Landing concrete jacking-pipes in the pit bottom

D) Building rings from a platform within a shaft

23.89

Oxygen cylinders should not be allowed to come into contact with which of the following substances?

A) Mud

B) Grease

C) Paint

D) Air

23.90

While walking through the tunnel you see a tear in the ventilation ducting. What should you do?

A) Report it to your supervisor

B) Try to repair it

C) Check if it has got any bigger at the end of the shift

D) Evacuate the tunnel

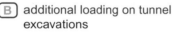

F
23

Answers: 23.85 = D   23.86 = D   23.87 = A   23.88 = D   23.89 = B   23.90 = A

23.91

If a blockage occurs during sprayed concrete operations the first action to be taken is to:

- A release pressure in the line
- B uncouple jointed pipes
- C increase pressure in the line
- D hit the line with a lump hammer

23.92

The most significant hazard from freshly applied sprayed concrete lining is:

- A material rebounding off the tunnel wall
- B material falling to crush workers
- C not wearing the right personal protective equipment (PPE)
- D workers being trapped by the equipment

23.93

One of the main hazards during shaft sinking operations is:

- A suspended loads
- B caulking
- C locomotive movements
- D maintenance work

23.94

The biggest risk from suspended loads during shaft sinking operations is:

- A damage to the shaft walls
- B falling objects and failure of the suspended load
- C restriction on the lifting capacity
- D restricted vision of the crane driver

23.95

When uncoupling/disconnecting hoses or pipework, operatives should firstly:

- A isolate/release any stored energy or pressure
- B check fluid levels
- C part uncouple/disconnect and check/listen for leaks
- D ensure that any stored energy is not released

23.96

What hazards need to be controlled during sprayed concrete operations?

- A Manual handling activities
- B Rebound of materials
- C Slips, trips and falls
- D All of these answers

F
23

23.97

Guards on grout mixers help prevent injury to operators. Before removing a guard you must ensure that the:

 A Health and Safety Executive (HSE) has been notified

B power has been isolated

C hose whip restraints are in place

D CDM co-ordinator has been notified

23.98

A grout gun inserted into a segment grout hole may present which of the following hazards?

A Shrinking grout hose

B Falls from height

C Blowout at injection point

D Hand-arm vibration

23.99

Which of the following personal protective equipment (PPE) is NOT normally required for robotic sprayed-concrete lining operations in tunnelling?

A Eye protection

B Respiratory protective equipment

C Disposable overalls

D Safety harness

F
23

24.1

You have to drill through a wall panel that you suspect contains an asbestos material. What should you do?

- A Use a low speed drill setting
- B Spray it with water as you drill
- C Put on an FFP3-rated dust mask before drilling
- D Stop work and report it

24.2

When a new piece of plant has been installed but has not been commissioned, how should it be left?

- A With all valves and switches turned off
- B With all valves and switches clearly labelled
- C With all valves and switches 'locked off'
- D With all valves and switches turned on and ready to use

24.3

Who can change a gas valve on a gas boiler?

- A A skilled engineer
- B A pipefitter
- C A registered Gas Safe engineer
- D Anybody

24.4

Who can solder a fitting on an isolated copper gas pipe?

- A A plumber
- B A pipefitter
- C A skilled welder
- D A Gas Safe registered engineer

24.5

When working in a riser, how should access be controlled?

- A By a site security operative
- B By those who are working in it
- C By the main contractor
- D By a permit to work system

24.6

If you find a coloured wire sticking out of an electrical plug what is the correct action to take?

- A Push it back into the plug and carry on working
- B Pull the wire clear of the plug and report it to your supervisor
- C Mark the item as defective and follow your company procedure for defective items
- D Take the plug apart and carry out a repair

F 24

24.7

Extension leads in use on a site should be:

A located so as to prevent a tripping hazard

B laid out in the shortest, most convenient route

C coiled on a drum or cable tidy

D raised on bricks

24.8

What should you do if you need additional temporary wiring for your power tools whilst working on site?

A Find some cable and extend the wiring yourself

B Stop work until an authorised supply has been installed

C Speak to an electrician and ask them to do the temporary wiring

D Disconnect a longer cable serving somewhere else and reconnect it to where you need it

24.9

Which of the following electrical equipment does NOT require portable appliance testing?

A Battery-powered rechargeable drill

B 110 volt electrical drill

C 110 volt portable halogen light

D Electric kettle

24.10

Temporary continuity bonding is carried out before removing and replacing sections of metallic pipework to:

A provide a continuous earth for the pipework installation

B prevent any chance of blowing a fuse

C maintain the live supply to the electrical circuit

D prevent any chance of corrosion to the pipework

24.11

Which type of power drill is most suitable for fixing a run of pipework outside in wet weather?

A Battery-powered drill

B Drill with 110 volt power supply

C Drill with 240 volt power supply

D Any mains voltage drill with a power breaker

24.12

What would you use to find out whether a wall into which you are about to drill contains an electric supply?

A A neon screwdriver

B A cable tracer

C A multimeter

D A hammer and chisel

F
24

Answers: 24.7 = A   24.8 = B   24.9 = A   24.10 = A   24.11 = A   24.12 = B

24.13

Where should liquefied petroleum gas (LPG) cylinders be positioned when supplying an appliance in a site cabin?

A) Inside the site cabin in a locked cupboard

B) Under the cabin

C) Inside the cabin next to the appliance

D) Outside the cabin

24.14

How should you position the exhaust of an engine-driven generator that has to be run inside a building?

A) Position the exhaust outside the building

B) Position the exhaust in a stairwell

C) Hang the exhaust in another room

D) Position the exhaust in a riser

24.15

How should cylinders containing liquefied petroleum gas (LPG) be stored on site?

A) In a locked cellar with clear warning signs

B) In a locked external compound at least 3 m from any oxygen cylinders

C) As close to the point of use as possible

D) Covered by a tarpaulin to shield the compressed cylinder from sunlight

24.16

You spill some oil on the floor and you do not have any absorbent material to clean the area. What should you do?

A) Spread it about to lessen the depth

B) Keep people out of the area and inform your supervisor

C) Do nothing, it will eventually soak into the floor

D) Warn other people as they tread through it

24.17

If you find a dangerous gas fitting that is likely to cause a death or specified injury, to whom MUST a formal report be sent?

A) The client

B) The gas board

C) The health and safety manager

D) The Health and Safety Executive (HSE)

24.18

What is the colour of propane gas cylinders?

A) Black

B) Maroon

C) Red/orange

D) Blue

F
24

Answers: 24.13 = D   24.14 = A   24.15 = B   24.16 = B   24.17 = D   24.18 = C

24.19

When drilling a hole through a wall you need to wear eye protection:

A when drilling overhead only

B when the drill bit exceeds 20 mm

C always, whatever the circumstances

D when drilling through concrete only

24.20

What personal protective equipment (PPE) should you wear when using a hammer drill to drill a 100 mm diameter hole through a brick wall?

A Gloves, breathing apparatus and boots

B Ear defenders, face mask and boots

C Ear defenders, breathing apparatus and barrier cream

D Barrier cream, boots and face mask

24.21

When using pipe-freezing equipment to isolate the damaged section of pipe, you should:

A always work in pairs when using pipe-freezing equipment

B never allow the freezing gas to come into direct contact with surface water

C never use pipe-freezing equipment on plastic pipe

D wear gloves to avoid direct contact with the skin and read the COSHH assessment

24.22

If you are working where welding is being carried out, what should be provided to protect you from 'welding flash'?

A Fire extinguishers

B Warning notices

C Screens

D Hi-vis vest

24.23

When using a blowtorch to joint copper tube and fittings in a domestic property, a fire extinguisher should be:

A available in the immediate work area

B held over the joint while you are using the blowtorch

C used to cool the fitting

D available only if a property is occupied

24.24

When using a blowtorch, you should stop using the blowtorch:

A immediately before leaving the job

B at least one hour before leaving the job

C at least two hours before leaving the job

D at least four hours before leaving the job

F
24

Answers: 24.19 = C   24.20 = B   24.21 = D   24.22 = C   24.23 = A   24.24 = B

24.25

When using a blowtorch near to flexible pipe lagging, you should:

(A) just remove enough lagging to carry out the job

(B) wet the lagging but leave it in place

(C) remove the lagging at least 3 m either side of the work

(D) remove the lagging at least 1 m either side of the work

24.26

When using a blowtorch near to timber, you should:

(A) carry out the work, taking care not to set fire to the timber

(B) wet the timber first and have a bucket of water handy

(C) use a non-combustible mat and have a fire extinguisher ready

(D) point the flame away from the timber and have a bucket of sand ready to put out the fire

24.27

The legionella bacteria that cause legionnaires' disease are most likely to be found in which of the following?

(A) A boiler operating at a temperature of 80°C

(B) An infrequently used shower hose outlet

(C) A cold water storage cistern containing water at 10°C

(D) A toilet pan

24.28

How are legionella bacteria passed on to humans?

(A) Through fine water droplets, such as sprays or mists

(B) By drinking dirty water

(C) Through contact with the skin

(D) From other people when they sneeze

24.29

When planning a lifting operation the sequence of operations to enable a lift to be carried out safely should be confirmed in:

(A) verbal instructions

(B) a method statement

(C) a radio telephone message

(D) a notice in the canteen

24.30

The SWL of lifting equipment is:

(A) never marked on the equipment but kept with the test certificates

(B) provided for guidance only

(C) may be exceeded by no more than 25%

(D) the absolute maximum safe working load

F
24

24.31

What must be clearly marked on all lifting equipment?

A Name of the manufacturer

B Safe working load

C Next test date

D Specification of material from which it's made

24.32

You are required to take up a length of floorboard to install pipework. Which of the following tools should you use?

A A hammer and wood chisel

B A hammer and screwdriver

C A hammer and bolster

D A chainsaw

24.33

What is the safest method of taking long lengths of copper pipe by van?

A Tying the pipes to the roof with copper wire

B Someone holding the pipes on the roof rack as you drive along

C Putting the pipes inside the van with the ends out of the passenger window

D Using a pipe rack fixed to the roof of the van

24.34

You are asked to move a cast-iron boiler some distance. What is the safest method?

A Get a workmate to carry it with you

B Drag it

C Roll it end-over-end

D Use a trolley or other manual handling aid

24.35

During a job you may need to work below a ground-level suspended timber floor. What is the first question you should ask?

A Can the work be performed from outside?

B Will temporary lighting be used?

C What is contained under the floor?

D How many ways in or out are there?

24.36

When carrying a ladder on a vehicle, what is the correct way of securing the ladder to the roof rack?

A Rope

B Bungee elastics

C Ladder clamps

D All of these answers

F
24

Answers: 24.31 = B   24.32 = C   24.33 = D   24.34 = D   24.35 = A   24.36 = C

24.37

Before using a ladder you must make sure that:

A it is secured to prevent it from moving sideways or sliding outwards

B no-one else has booked the ladder for their work

C an apprentice or workmate is standing by in case you slip and fall

D the weather forecast is for a bright, clear day

24.38

When positioning and erecting a stepladder, which of the following is essential for its safe use?

A It has a tool tray towards the top of the steps

B The restraint mechanism is spread to its full extent

C You will be able to reach the job by standing on the top step

D A competent person has positioned and erected the steps

24.39

Generally, how many working platforms should be in use at any one time on a mobile tower?

A One

B Two

C Three

D Four

24.40

What is the recommended maximum height for a free-standing mobile tower?

A There is no restriction

B 2 m

C In accordance with the manufacturer's recommendations

D 12 m

24.41

What is the first thing you should do after getting on to the platform of a mobile tower?

A Check that the brakes are locked on

B Check the mobile tower to make sure that it has been correctly assembled

C Close the access hatch to prevent falls of personnel, tools or equipment

D Make sure that the tower does not rock or wobble

24.42

What should be done before a mobile tower is moved?

A All people and equipment must be removed from the platform

B A permit to work is required

C The principal contractor must give their approval

D Arrangements must be made with the forklift truck driver

F
24

24.43

What must be done first before any roof work is carried out?

- A A risk assessment must be carried out
- B The operatives working on the roof must be trained in the use of safety harnesses
- C Permits to work must be issued only to those allowed to work on the roof
- D A weather forecast must be obtained

24.44

What is edge protection designed to do?

- A Make access to the roof easier
- B Secure tools and materials close to the edge
- C Prevent rainwater running off the roof onto workers below
- D Prevent the fall of people and materials

24.45

What is the first thing you should do after getting on to the platform of a correctly erected mobile tower?

- A Check that the brakes are locked on
- B Check the mobile tower to make sure that it has been correctly assembled
- C Close the access hatch to prevent falls of personnel, tools or equipment
- D Make sure that the tower does not rock or wobble

24.46

When assembling a mobile tower what major overhead hazard must you be aware of?

- A Water pipes
- B Cable trays
- C False ceilings
- D Suspended electric cables

24.47

What should folding stepladders be used for?

- A General access on site
- B Short-term activities lasting minutes that don't involve stretching
- C All site activities where a straight ladder cannot be used
- D Getting on and off mobile towers

24.48

When drilling a hole for a boiler flue, from the outside of a property at first floor level, which of the following means of access should you use?

- A A long ladder
- B Borrow some scaffolding and erect it
- C A mobile tower
- D Packing cases to stand on

F
24

24.49

When working on a roof to install a flexible flue liner into an existing chimney, you should:

A work from a roof ladder securely hooked over the ridge

B use a stable working platform, with handrails, around or next to the chimney

C scramble up the roof tiles to get to the chimney

D get your mate to do the job while you hold a rope tied to them

24.50

Stepladders must only be used:

A inside buildings

B if they are in good condition and suitable

C if they are made of aluminium

D if they are less than 1.75 m high

F
24

25.1

When a new piece of plant has been installed but has not been commissioned, how should it be left?

A) With all valves and switches turned off

B) With all valves and switches clearly labelled

C) With all valves and switches 'locked off'

D) With all valves and switches turned on and ready to use

25.2

Who is allowed to install natural gas pipework?

A) A skilled engineer

B) A pipefitter

C) A registered Gas Safe engineer

D) Anybody

25.3

Who should carry out pressure testing on pipework or vessels?

A) Anyone who is available

B) A competent person

C) A Health and Safety Executive (HSE) inspector

D) A building control officer

25.4

When working in a riser, how should access be controlled?

A) By a site security operative

B) By those who are working in it

C) By the main contractor

D) By a permit to work system

25.5

While working on your own and tracing pipework in a building, the pipes enter a service duct. You should:

A) go into the service duct and continue to trace the pipework

B) ask someone in the building to act as your second person

C) put on your personal protective equipment (PPE) and carry on with the job

D) stop work until a risk assessment has been carried out

25.6

If you find a coloured wire sticking out of an electrical plug what is the correct action to take?

A) Push it back into the plug and carry on working

B) Pull the wire clear of the plug and report it to your supervisor

C) Mark the item as defective and follow your company procedure for defective items

D) Take the plug apart and carry out a repair

F
25

25.7

Extension leads in use on a site should be:

A located so as to prevent a tripping hazard

B laid out in the shortest, most convenient route

C coiled on a drum or cable tidy

D raised on bricks

25.8

What should you do if you need additional temporary wiring for your power tools whilst working on site?

A Find some cable and extend the wiring yourself

B Stop work until an authorised supply has been installed

C Speak to an electrician and ask them to do the temporary wiring

D Disconnect a longer cable serving somewhere else and reconnect it to where you need it

25.9

Which of the following electrical equipment does NOT require portable appliance testing?

A Battery-powered rechargeable drill

B 110 volt electrical drill

C 110 volt portable halogen light

D Electric kettle

25.10

Where should liquefied petroleum gas (LPG) cylinders be positioned when supplying an appliance in a site cabin?

A Inside the site cabin in a locked cupboard

B Under the cabin

C Inside the cabin next to the appliance

D Outside the cabin

25.11

How should you position the exhaust of an engine-driven generator that has to be run inside a building?

A Position the exhaust outside the building

B Position the exhaust in a stairwell

C Hang the exhaust in another room

D Position the exhaust in a riser

25.12

How should cylinders containing liquefied petroleum gas (LPG) be stored on site?

A In a locked cellar with clear warning signs

B In a locked external compound at least 3 m from any oxygen cylinders

C As close to the point of use as possible

D Covered by a tarpaulin to shield the compressed cylinder from sunlight

F
25

25.13

You spill some oil on the floor and you do not have any absorbent material to clean the area. What should you do?

- A Spread it about to lessen the depth
- B Keep people out of the area and inform your supervisor
- C Do nothing, it will eventually soak into the floor
- D Warn other people as they tread through it

25.14

What guarding is required when a pipe threading machine is in use?

- A A length of red material hung from the exposed end of the pipe
- B A barrier at the exposed end of the pipe only
- C A barrier around the whole length of the pipe
- D Warning notices in the work area

25.15

When using pipe-freezing equipment to isolate the damaged section of pipe, you should:

- A always work in pairs
- B never allow the freezing gas to come into direct contact with surface water
- C never use pipe-freezing equipment on plastic pipe
- D wear gloves to avoid direct contact with the skin and read the COSHH assessment

25.16

Why is it important to know the difference between propane and butane equipment?

- A Propane equipment operates at higher pressure
- B Propane equipment operates at lower pressure
- C Propane equipment is cheaper
- D Propane equipment can be used with smaller, easy-to-handle cylinders

25.17

Which of the following statements is true?

- A Both propane and butane are heavier than air
- B Butane is heavier than air while propane is lighter than air
- C Propane is heavier than air while butane is lighter than air
- D Both propane and butane are lighter than air

F
25

25.18

Apart from the cylinders used in gas-powered forklift trucks, you should never see liquefied petroleum gas (LPG) cylinders placed on their sides during use because:

A) it would give a faulty reading on the contents gauge, resulting in flashback

B) air could be drawn into the cylinder, creating a dangerous mixture of gases

C) the liquid gas would be at too low a level to allow the torch to burn correctly

D) the liquid gas could be drawn from the cylinder, creating a safety hazard

25.19

What is the method of checking for leaks after connecting a liquefied petroleum gas (LPG) regulator to the bottle?

A) Test with a lighted match

B) Sniff the connections to detect the smell of gas

C) Listen to hear for escaping gas

D) Apply leak detection fluid to the connections

25.20

What is the most likely risk of injury when cutting a pipe with hand-operated pipe cutters?

A) Your fingers may become trapped between the cutting wheel and the pipe

B) The inside edge of the cut pipe becomes extremely sharp to touch

C) Continued use can cause muscle damage

D) Pieces of sharp metal could fly off

25.21

Why is it essential to take great care when handling oxygen cylinders?

A) They contain highly flammable compressed gas

B) They contain highly flammable liquid gas

C) They are filled to extremely high pressures

D) They contain poisonous gas

25.22

When drilling a hole through a wall you need to wear eye protection:

A) when drilling overhead only

B) when the drill bit exceeds 20 mm

C) always, whatever the circumstances

D) when drilling through concrete only

F
25

25.23

If you are working where welding is being carried out, what should be provided to protect you from 'welding flash'?

- A) Fire extinguishers
- B) Warning notices
- C) Screens
- D) Hi-vis vest

25.24

What is the main hazard associated with flame-cutting and welding?

- A) Gas poisoning
- B) Fire
- C) Dropping a gas cylinder
- D) Not having a 'hot work' permit

25.25

When using a blowtorch, you should:

- A) stop using the blowtorch immediately before leaving the job
- B) stop using the blowtorch at least one hour before leaving the job
- C) stop using the blowtorch at least two hours before leaving the job
- D) stop using the blowtorch at least four hours before leaving the job

25.26

When using a blowtorch near to flexible pipe lagging, you should:

- A) remove the lagging at least 1 m either side of the work
- B) just remove enough lagging to carry out the work
- C) remove the lagging at least 3 m either side of the work
- D) wet the lagging but leave it in place

25.27

When using a blowtorch near to timber, you should:

- A) carry out the work taking care not to catch the timber
- B) point the flame away from the timber and have a bucket of sand ready to put out the fire
- C) wet the timber first and keep a bucket of water handy
- D) use a non-combustible mat and have a fire extinguisher ready

25.28

What is the colour of an acetylene cylinder?

- A) Orange
- B) Black
- C) Green
- D) Maroon

F
25

Answers: 25.23 = C   25.24 = B   25.25 = B   25.26 = A   25.27 = D   25.28 = D

25.29

What particular item of personal protective equipment (PPE) should you use when oxyacetylene brazing?

- A. Ear defenders
- B. Clear goggles
- C. Green-tinted goggles
- D. Dust mask

25.30

When using oxyacetylene brazing equipment, the bottles should be:

- A. laid on their side and secured
- B. stood upright and secured
- C. stood upside down
- D. angled at 45° and secured against falling

25.31

The use of oxyacetylene equipment is NOT recommended for which of the following jointing methods?

- A. Jointing copper pipe using hard soldering
- B. Jointing copper tube using capillary soldered fittings
- C. Jointing mild steel tube
- D. Jointing sheet lead

25.32

Where should acetylene gas-welding bottles be stored when they are not in use?

- A. Outside in a special storage compound
- B. In a special rack in a company van
- C. Inside a building in a locked cupboard
- D. With oxygen bottles

25.33

When planning a lifting operation the sequence of operations to enable a lift to be carried out safely should be confirmed in:

- A. verbal instructions
- B. a method statement
- C. a radio telephone message
- D. a notice in the canteen

25.34

The SWL of lifting equipment is:

- A. never marked on the equipment but kept with the test certificates
- B. provided for guidance only
- C. may be exceeded by no more than 25%
- D. the absolute maximum safe working load

F
25

25.35

What must be clearly marked on all lifting equipment?

 A. Name of the manufacturer

B. Safe working load

C. Next test date

D. Specification of material from which it's made

25.36

Which TWO of the following are essential safety checks to be carried out before using oxyacetylene equipment?

 A. The cylinders are full

B. The cylinders, hoses and flashback arresters are in good condition

C. The trolley wheels are the right size

D. The area is well ventilated and clear of any obstructions

E. The cylinders are the right weight

25.37

During the pressure testing of pipework or vessels, who should be present?

A. The architect

B. The site foreman

C. Only those involved in carrying out the test

D. Anybody

25.38

When using an electrically powered threading machine you should make sure that:

A. the power supply is 24 volts

B. the power supply is 415 volts and the machine is fitted with a guard

C. your clothing cannot get caught on rotating parts of the machine

D. the machine is only used in your compound

25.39

Before using a ladder you must make sure that:

A. it is secured to prevent it from moving sideways or sliding outwards

B. no-one else has booked the ladder for their work

C. an apprentice or workmate is standing by in case you slip and fall

D. the weather forecast is for a bright, clear day

25.40

When positioning and erecting a stepladder, which of the following is essential for its safe use?

A. It has a tool tray towards the top of the steps

B. The restraint mechanism is spread to its full extent

C. You will be able to reach the job by standing on the top step

D. A competent person has positioned and erected the steps

F
25

25.41

Generally, how many working platforms should be in use at any one time on a mobile tower?

- [A] One
- [B] Two
- [C] Three
- [D] Four

25.42

What is the recommended maximum height for a free-standing mobile tower?

- [A] There is no restriction
- [B] 2 m
- [C] In accordance with the manufacturer's recommendations
- [D] 12 m

25.43

What is the first thing you should do after getting on to the platform of a mobile tower?

- [A] Check that the brakes are locked on
- [B] Check the mobile tower to make sure that it has been correctly assembled
- [C] Close the access hatch to prevent falls of personnel, tools or equipment
- [D] Make sure that the tower does not rock or wobble

25.44

What should be done before a mobile tower is moved?

- [A] All people and equipment must be removed from the platform
- [B] A permit to work is required
- [C] The principal contractor must give their approval
- [D] Arrangements must be made with the forklift truck driver

25.45

What must be done first before any roof work is carried out?

- [A] A risk assessment must be carried out
- [B] The operatives working on the roof must be trained in the use of safety harnesses
- [C] Permits to work must be issued only to those allowed to work on the roof
- [D] A weather forecast must be obtained

25.46

What is edge protection designed to do?

- [A] Make access to the roof easier
- [B] Secure tools and materials close to the edge
- [C] Prevent rainwater running off the roof onto workers below
- [D] Prevent the fall of people and materials

F
25

25.47

What is the first thing you should do after getting on to the platform of a correctly erected mobile tower?

A Check that the brakes are locked on

B Check the mobile tower to make sure that it has been correctly assembled

C Close the access hatch to prevent falls of personnel, tools or equipment

D Make sure that the tower does not rock or wobble

25.48

When assembling a mobile tower what major overhead hazard must you be aware of?

A Water pipes

B Cable trays

C False ceilings

D Suspended electric cables

25.49

What should folding stepladders be used for?

A General access on site

B Short-term activities lasting minutes that don't involve stretching

C All site activities where a straight ladder cannot be used

D Getting on and off mobile towers

25.50

You are asked to install high-level ductwork from a platform that has no edge protection and is located above an open stairwell. You should:

A get on with the job, but keep away from the edge of the platform

B not start work until your work platform has been fitted with guard-rails and toe-boards

C get on with the job, ensuring that a workmate stays close by

D get on with the job, provided that if you fall the stairwell guard-rail will prevent you from falling further

F
25

26.1

You are working on a refurbishment and removing some old ductwork when you notice that it has been insulated using a white, powdery material that could be asbestos. What should you do?

A Remove the insulation while using a vacuum cleaner

B Get an FFP3-rated dust mask for yourself and get some assistance to do the job quickly

C Stop work immediately and report it

D Take off the material carefully and place it in a sealed container

26.2

When a new piece of plant has been installed but has not been commissioned, how should it be left?

A With all valves and switches turned off

B With all valves and switches clearly labelled

C With all valves and switches 'locked off'

D With all valves and switches turned on and ready to use

26.3

Who should carry out leakage testing of a newly installed ductwork system?

A The installation contractor

B The property owner

C The designer

D A trained and competent person

26.4

When working in a riser, how should access be controlled?

A By a site security operative

B By those who are working in it

C By the main contractor

D By a permit to work system

26.5

If you find a coloured wire sticking out of an electrical plug what is the correct action to take?

A Push it back into the plug and carry on working

B Pull the wire clear of the plug and report it to your supervisor

C Mark the item as defective and follow your company procedure for defective items

D Take the plug apart and carry out a repair

26.6

Extension leads in use on a site should be:

A located so as to prevent a tripping hazard

B laid out in the shortest, most convenient route

C coiled on a drum or cable tidy

D raised on bricks

F
26

Answers: 26.1 = C   26.2 = C   26.3 = D   26.4 = D   26.5 = C   26.6 = A

26.7

What should you do if you need additional temporary wiring for your power tools whilst working on site?

A Find some cable and extend the wiring yourself

B Stop work until an authorised supply has been installed

C Speak to an electrician and ask them to do the temporary wiring

D Disconnect a longer cable serving somewhere else and reconnect it to where you need it

26.8

Which of the following electrical equipment does NOT require portable appliance testing?

A Battery-powered rechargeable drill

B 110 volt electrical drill

C 110 volt portable halogen light

D Electric kettle

26.9

Where should liquefied petroleum gas (LPG) cylinders be positioned when supplying an appliance in a site cabin?

A Inside the site cabin in a locked cupboard

B Under the cabin

C Inside the cabin next to the appliance

D Outside the cabin

26.10

How should you position the exhaust of an engine-driven generator that has to be run inside a building?

A Position the exhaust outside the building

B Position the exhaust in a stairwell

C Hang the exhaust in another room

D Position the exhaust in a riser

26.11

How should cylinders containing liquefied petroleum gas (LPG) be stored on site?

A In a locked cellar with clear warning signs

B In a locked external compound at least 3 m from any oxygen cylinders

C As close to the point of use as possible

D Covered by a tarpaulin to shield the compressed cylinder from sunlight

26.12

You spill some oil on the floor and you do not have any absorbent material to clean the area. What should you do?

A Spread it about to lessen the depth

B Keep people out of the area and inform your supervisor

C Do nothing, it will eventually soak into the floor

D Warn other people as they tread through it

F
26

26.13

What should you do if you see the side of an abrasive disc on a disc cutter being used to grind down the ends of a drop rod?

A | Check that the correct type of disc is being used

B | Stop the person immediately

C | Go and find your supervisor and report it

D | Check that the disc has not been subjected to over-speeding

26.14

You have to cut some flexible aluminium ductwork that is pre-insulated with a fibreglass material. Which TWO of the following should you use?

A | A hacksaw

B | Respiratory protection

C | A disc cutter

D | A set of tin snips

E | Ear defenders

26.15

A person who has been using a solvent-based ductwork sealant is complaining of headaches and feeling sick. What is the FIRST thing you should do?

A | Let them carry on working but try to keep a close watch on them

B | Get them a drink of water and a headache tablet

C | Get them out to fresh air and make them rest

D | Stop work and tell your supervisor

26.16

What additional control measure must be put in place when welding in-situ galvanised ductwork?

A | Screens

B | Fume extraction

C | Warning signs

D | Hearing protection

26.17

When jointing plastic-coated metal ductwork, which of the following methods of jointing presents the most serious risk to health?

A | Welding

B | Taping

C | Riveting

D | Nuts and bolts

26.18

You are removing a run of ductwork in an unoccupied building and notice a hypodermic syringe behind it. What should you do?

A | Ensure the syringe is empty, remove the syringe and place it with the rubbish

B | Wear gloves, break the syringe into small pieces and flush it down the drain

C | Notify the supervisor, cordon off the area and call the emergency services

D | Wear gloves, use grips to remove the syringe to a safe place and report your find

F
26

Answers: 26.13 = B   26.14 = B, D   26.15 = C   26.16 = B   26.17 = A   26.18 = D

26.19

If you are working where welding is being carried out, what should be provided to protect you from 'welding flash'?

 A Fire extinguishers

B Warning notices

C Screens

D Hi-vis vest

26.20

When planning a lifting operation the sequence of operations to enable a lift to be carried out safely should be confirmed in:

A verbal instructions

B a method statement

C a radio telephone message

D a notice in the canteen

26.21

The SWL of lifting equipment is:

A never marked on the equipment but kept with the test certificates

B provided for guidance only

C may be exceeded by no more than 25%

D the absolute maximum safe working load

26.22

What must be clearly marked on all lifting equipment?

A Name of the manufacturer

B Safe working load

C Next test date

D Specification of material from which it's made

26.23

When using a material hoist you notice that the lifting cable is frayed. What should you do?

A Get the job done as quickly as possible

B Straighten out the cable using mole grips

C Do not use the hoist and report the problem

D Be very careful when using the hoist

26.24

When drilling a hole through a wall you need to wear eye protection:

A when drilling overhead only

B when the drill bit exceeds 20 mm

C always, whatever the circumstances

D when drilling through concrete only

F
26

26.25

In addition to a safety helmet and protective footwear, what personal protective equipment (PPE) should you wear when using a hammer drill?

A Gloves and breathing apparatus

B Hearing protection, face mask and eye protection

C Hearing protection, breathing apparatus and barrier cream

D Barrier cream and face mask

26.26

How should you leave the ends of ductwork after using a solvent-based sealant?

A Seal up all the open ends to ensure that dirt cannot get into the system

B Ensure that the lids are left off tins of solvent

C Remove any safety signs or notices

D Leave inspection covers off and erect 'No smoking' signs

26.27

Before taking down a run of ductwork, what is the first thing you should do?

A Get a big skip to put it in

B Cut through the support rods

C Clean the ductwork to remove all dust

D Assess the task to be undertaken

26.28

You are asked to move a fan-coil unit some distance. What is the safest way to do it?

A Get a workmate to carry it with you

B Drag it

C Roll it end-over-end

D Use a trolley or other manual handling aid

26.29

While fitting a fire damper into a ductwork system you notice that, due to a manufacturing fault, it may not operate properly. You should:

A install it anyway, as it is

B fix it so that it stays open, and then install it

C not fit the damper and report the fault

D leave it out of the ductwork system altogether

26.30

You have to dismantle some waste-extract ductwork. What is the first thing you should do?

A Arrange for a skip to put it in

B Find out what the ductwork may be contaminated with

C Check that the duct supports are strong enough to cope with the dismantling

D Make sure there are enough disc cutters to do the job

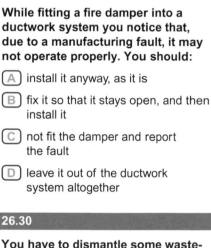

F
26

26.31

While using a 'genie' hoist you notice that part of the hoist is buckling slightly. You should:

- [A] lower the load immediately
- [B] carry on with the job, while keeping an eye on the buckling metal
- [C] straighten out the buckled metal and then get on with the lifting operation
- [D] get the job finished quickly

26.32

When carrying out solvent welding on plastic ductwork, what particular safety measure must be applied?

- [A] The area must be well ventilated
- [B] The supervisor must be present
- [C] A hard hat must be worn
- [D] It must be done in daylight

26.33

Which of the following need you NOT do before using a cleaning agent or biocide in a ductwork system?

- [A] Ask for advice from the cleaning agent or biocide manufacturer
- [B] Read the COSHH assessment for the material, carrying out a risk assessment and producing a method statement for the work
- [C] Consult the building occupier
- [D] Check what the ductwork will carry in the future

26.34

Which of the following need NOT be done before cleaning a system in industrial, laboratory or other premises where you might encounter harmful particulates?

- [A] Examine the system
- [B] Collect a sample from the ductwork
- [C] Run the system under overload conditions
- [D] Prepare a job-specific risk assessment and method statement

26.35

Where it is necessary to enter ductwork, which are the TWO main factors that need to be considered?

- [A] Working in a confined space
- [B] What the ductwork will carry in the future
- [C] The cleanliness of the ductwork
- [D] Wearing kneepads
- [E] The strength of the ductwork and its supports

26.36

What particular factor should be considered before working on a kitchen extraction system?

- [A] Access to the ductwork
- [B] The cooking deposits within the ductwork
- [C] The effect on the future performance of the system
- [D] The effect on food preparation

F
26

26.37

What should you do before painting the external surface of ductwork?

A) Clean the paintbrushes

B) Read the COSHH assessment before using the paint

C) Switch off the system

D) Put on eye protection

26.38

Before using a ladder you must make sure that:

A) it is secured to prevent it from moving sideways or sliding outwards

B) no-one else has booked the ladder for their work

C) an apprentice or workmate is standing by in case you slip and fall

D) the weather forecast is for a bright, clear day

26.39

When positioning and erecting a stepladder, which of the following is essential for its safe use?

A) It has a tool tray towards the top of the steps

B) The restraint mechanism is spread to its full extent

C) You will be able to reach the job by standing on the top step

D) A competent person has positioned and erected the steps

26.40

Generally, how many working platforms should be in use at any one time on a mobile tower?

A) One

B) Two

C) Three

D) Four

26.41

What is the recommended maximum height for a free-standing mobile tower?

A) There is no restriction

B) 2 m

C) In accordance with the manufacturer's recommendations

D) 12 m

26.42

What is the first thing you should do after getting on to the platform of a mobile tower?

A) Check that the brakes are locked on

B) Check the mobile tower to make sure that it has been correctly assembled

C) Close the access hatch to prevent falls of personnel, tools or equipment

D) Make sure that the tower does not rock or wobble

F 26

26.43

What should be done before a mobile tower is moved?

A | All people and equipment must be removed from the platform

B | A permit to work is required

C | The principal contractor must give their approval

D | Arrangements must be made with the forklift truck driver

26.44

What must be done first before any roof work is carried out?

A | A risk assessment must be carried out

B | The operatives working on the roof must be trained in the use of safety harnesses

C | Permits to work must be issued only to those allowed to work on the roof

D | A weather forecast must be obtained

26.45

What is edge protection designed to do?

A | Make access to the roof easier

B | Secure tools and materials close to the edge

C | Prevent rainwater running off the roof onto workers below

D | Prevent the fall of people and materials

26.46

What is the first thing you should do after getting on to the platform of a correctly erected mobile tower?

A | Check that the brakes are locked on

B | Check the mobile tower to make sure that it has been correctly assembled

C | Close the access hatch to prevent falls of personnel, tools or equipment

D | Make sure that the tower does not rock or wobble

26.47

When assembling a mobile tower what major overhead hazard must you be aware of?

A | Water pipes

B | Cable trays

C | False ceilings

D | Suspended electric cables

26.48

What should folding stepladders be used for?

A | General access on site

B | Short-term activities lasting minutes that don't involve stretching

C | All site activities where a straight ladder cannot be used

D | Getting on and off mobile towers

F
26

26.49

You are asked to install high-level ductwork from a platform that has no edge protection and is located above an open stairwell. You should:

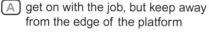 get on with the job, but keep away from the edge of the platform

B not start work until your work platform has been fitted with guard-rails and toe-boards

C get on with the job, ensuring that a workmate stays close by

D get on with the job, provided that if you fall the stairwell guard-rail will prevent you from falling further

26.50

You have been asked to install a run of ceiling-mounted ductwork across a large open-plan area, which has a good floor. It is best to do this:

A using stepladders

B using scaffold boards and floor stands

 using packing cases to stand on

D from mobile access towers fitted with guard-rails and toe-boards

26.51

You have to carry out a job over a few days on the flat roof of a two-storey building, about 1 m from the edge of the roof, which has a very low parapet. You should:

A carry on with the job, provided that you don't get dizzy with heights

B use a full body harness, lanyard and anchor while doing the job

C ask for double guard-rails and a toe-board to be installed to prevent you falling

D get your mate to do the work, while you hold on to them

F
26

27.1

When a new piece of plant has been installed but has not been commissioned, how should it be left?

A With all valves and switches turned off

B With all valves and switches clearly labelled

C With all valves and switches 'locked off'

D With all valves and switches turned on and ready to use

27.2

When working on refrigeration systems containing hydrocarbon (HC) gases, what particular danger needs to be considered?

A There should be no sources of ignition

B Special personal protective equipment (PPE) should be worn

C Extra lighting is needed

D The work cannot be carried out when the weather is hot

27.3

When it is necessary to cut into an existing refrigerant pipe, should you:

A vent the gas in the pipework to atmosphere

B recover the refrigerant gas and make a record of it, then do the work

C work on the pipework with the refrigerant gas still in it

D not carry out the work at all, because of the risks

27.4

What is the FIRST thing that should be done when a new refrigeration system has been installed?

A It should be pressure and leak tested

B It should be filled with refrigerant

C It should be left open to air

D It should be turned off at the electrical switch

27.5

On water-cooled systems, the water in the system should be:

A replaced annually

B chemically treated

C properly filtered

D drinking water

27.6

Who is permitted to install, service or maintain systems that contain or are designed to contain refrigerant gases?

A A registered Gas Safe engineer

B The person whose plant contains the gas

C A competent, trained person who works for an F-gas registered company

D A fully qualified electrician

F
27

27.7

When working in a riser, how should access be controlled?

A | By a site security operative

B | By those who are working in it

C | By the main contractor

D | By a permit to work system

27.8

If you find a coloured wire sticking out of an electrical plug what is the correct action to take?

A | Push it back into the plug and carry on working

B | Pull the wire clear of the plug and report it to your supervisor

C | Mark the item as defective and follow your company procedure for defective items

D | Take the plug apart and carry out a repair

27.9

Extension leads in use on a site should be:

A | located so as to prevent a tripping hazard

B | laid out in the shortest, most convenient route

C | coiled on a drum or cable tidy

D | raised on bricks

27.10

What should you do if you need additional temporary wiring for your power tools whilst working on site?

A | Find some cable and extend the wiring yourself

B | Stop work until an authorised supply has been installed

C | Speak to an electrician and ask them to do the temporary wiring

D | Disconnect a longer cable serving somewhere else and reconnect it to where you need it

27.11

Which of the following electrical equipment does NOT require portable appliance testing?

A | 110 volt electrical power tool

B | Battery-powered rechargeable power tool

C | 240 volt electrical power tool

D | 240 volt charger for battery-powered tools

27.12

When repairing an electrically driven compressor, what is the minimum safe method of isolation?

A | Pressing the stop button

B | Pressing the emergency stop button

C | Turning off the local isolator

D | Locking off and tagging out the local isolator

F
27

27.13

When a refrigerant leak is reported in a closed area, what should you do first before entering the area?

A Ventilate the area

B Establish that it is safe to enter

C Get a torch

D Wear safety footwear

27.14

Where should liquefied petroleum gas (LPG) cylinders be positioned when supplying an appliance in a site cabin?

A Inside the site cabin in a locked cupboard

B Under the cabin

C Inside the cabin next to the appliance

D Outside the cabin

27.15

How should you position the exhaust of an engine-driven generator that has to be run inside a building?

A Position the exhaust outside the building

B Position the exhaust in a stairwell

C Hang the exhaust in another room

D Position the exhaust in a riser

27.16

How should cylinders (full or empty) containing liquefied petroleum gas (LPG) or acetylene be stored on site?

A In a locked cellar with clear warning signs

B In a locked external compound at least 3 m from any oxygen cylinders

C As close to the point of use as possible

D Covered by a tarpaulin to shield the compressed cylinder from sunlight

27.17

You spill some oil on the floor and you do not have any absorbent material to clean the area. What should you do?

A Spread it about to lessen the depth

B Keep people out of the area and inform your supervisor

C Do nothing, it will eventually soak into the floor

D Warn other people as they tread through it

F
27

27.18

Which is the safest place to store refrigerant cylinders when they are not in use?

A Outside in a special locked storage compound

B In a company vehicle

C Inside the building in a locked cupboard

D In the immediate work area, ready for use the next day

27.19

If refrigerant gases are released into a closed room in a building they would:

A sink to the floor

B rise to the ceiling

C stay at the same level

D disperse safely within the room

27.20

You have to drill through a wall panel that you suspect contains an asbestos material. What should you do?

A Ignore it and carry on

B Put on safety goggles

C Put on a dust mask

D Stop work and report it

27.21

When using a van to transport a refrigerant bottle, how should it be carried?

A In the back of the van

B In the passenger footwell of the van

C In a purpose-built container within the van

D In the van with all the windows open

27.22

What safety devices should be fitted between the pipes and the gauges of oxy-propane brazing equipment?

A Non-return valves

B On-off taps

C Flame retardant tape

D Flashback arresters

27.23

Why is it essential to take great care when handling oxygen cylinders?

A They contain highly flammable compressed gas

B They contain highly flammable liquid gas

C They are filled to extremely high pressures

D They contain poisonous gas

F
27

27.24

What part of the body could suffer long-term damage when hand-bending copper pipe using a spring?

A. Elbows

B. Shoulders

C. Back

D. Knees

27.25

If you are working where welding is being carried out, what should be provided to protect you from 'welding flash'?

A. Fire extinguishers

B. Warning notices

C. Screens

D. Hi-vis vest

27.26

When using a blowtorch or brazing equipment to joint copper tube and fittings in a property, a fire extinguisher should be:

A. available in the immediate work area

B. held over the joint while you are using the blowtorch

C. used to cool the fitting

D. available only if a property is occupied

27.27

When undertaking hot works, such as using a blowtorch or brazing equipment, you should:

A. stop using the blowtorch or brazing equipment immediately before leaving the job

B. stop using the blowtorch or brazing equipment at least one hour before leaving the job

C. stop using the blowtorch or brazing equipment at least two hours before leaving the job

D. stop using the blowtorch or brazing equipment at least four hours before leaving the job

27.28

When pressure testing, oxygen must never be used because:

A. the molecules of the gas are too small

B. the pressure required would not be reached

C. when oxygen meets oil in a compressor it could explode and cause serious injury or death

D. there is too much temperature pressure difference and a true record will not be given

F 27

27.29

When pressure testing with nitrogen you should ensure that:

A the nitrogen bottle is laid down to avoid it falling over

B the nitrogen gauge can take the pressure required and that the bottle is secured upright

C the temperature of the bottle is at room temperature to avoid a temperature pressure difference

D purge brazing has taken place during the installation

27.30

When planning a lifting operation the sequence of operations to enable a lift to be carried out safely should be confirmed in:

A verbal instructions

B a method statement

C a radio telephone message

D a notice in the canteen

27.31

The SWL of lifting equipment is:

A never marked on the equipment but kept with the test certificates

B provided for guidance only

C may be exceeded by no more than 25%

D the absolute maximum safe working load

27.32

What must be clearly marked on all lifting equipment?

A Name of the manufacturer

B Safe working load

C Next test date

D Specification of material from which it's made

27.33

What particular item of personal protective equipment (PPE) should you use when oxyacetylene brazing?

A Ear defenders

B Clear goggles

C Green-tinted goggles

D Dust mask

27.34

When using oxyacetylene brazing equipment, the bottles should be:

A laid on their side

B stood upright and secured

C stood upside down

D angled at 45°

F
27

27.35

Which TWO of the following are essential safety checks that need to be carried out before using oxyacetylene equipment?

A The cylinders are full

B The cylinders, hoses and flashback arresters are in good condition

C The trolley wheels are the right size

D The area is well ventilated and clear of any obstructions

E The cylinders are the right weight

27.36

When handling refrigerant gases, what personal protective equipment (PPE) should you wear as a minimum?

A Eye protection, overalls, thermal-resistant gloves and helmet

B Eye protection, overalls, thermal-resistant gloves and safety boots

C Eye protection, overalls, harness and safety boots

D Overalls, thermal-resistant gloves, helmet and safety boots

27.37

What should you establish before entering a cold room?

A The size of the cold room

B The temperature of the cold room

C Whether the exit door is fitted with an internal handle

D Whether there are lights and power in the cold room

27.38

Before using a ladder you must make sure that:

A it is secured to prevent it from moving sideways or sliding outwards

B no-one else has booked the ladder for their work

C an apprentice or workmate is standing by in case you slip and fall

D the weather forecast is for a bright, clear day

27.39

When positioning and erecting a stepladder, which of the following is essential for its safe use?

A It has a tool tray towards the top of the steps

B The restraint mechanism is spread to its full extent

C You will be able to reach the job by standing on the top step

D A competent person has positioned and erected the steps

27.40

Generally, how many working platform levels should be used on a mobile tower at any one time, once it has been correctly erected?

A One

B Two

C Three

D Four

F
27

Answers: 27.35 = B, D   27.36 = B   27.37 = C   27.38 = A   27.39 = B   27.40 = A

27.41

What is the recommended maximum height for a free-standing mobile tower?

A There is no restriction

B 2 m

C In accordance with the manufacturer's recommendations

D 12 m

27.42

What is the first thing you should do after getting on to the platform of a mobile tower?

A Check that the brakes are locked on

B Check the mobile tower to make sure that it has been correctly assembled

C Close the access hatch to prevent falls of personnel, tools or equipment

D Make sure that the tower does not rock or wobble

27.43

What should be done before a mobile tower is moved?

A All people and equipment must be removed from the platform

B A permit to work is required

C The principal contractor must give their approval

D Arrangements must be made with the forklift truck driver

27.44

What must be done first before any roof work is carried out?

A A risk assessment must be carried out

B The operatives working on the roof must be trained in the use of safety harnesses

C Permits to work must be issued only to those allowed to work on the roof

D A weather forecast must be obtained

27.45

What is edge protection designed to do?

A Make access to the roof easier

B Secure tools and materials close to the edge

C Prevent rainwater running off the roof onto workers below

D Prevent the fall of people and materials

27.46

What is the first thing you should do after getting on to the platform of a correctly erected mobile tower?

A Check that the brakes are locked on

B Check the mobile tower to make sure that it has been correctly assembled

C Close the access hatch to prevent falls of personnel, tools or equipment

D Make sure that the tower does not rock or wobble

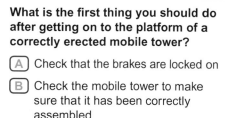

**F
27**

Answers: 27.41 = C   27.42 = C   27.43 = A   27.44 = A   27.45 = D   27.46 = C

27.47

When assembling a mobile tower what major overhead hazard must you be aware of?

 Water pipes

B Cable trays

C False ceilings

D Suspended electric cables

27.48

What should folding stepladders be used for?

 General access on site

B Short-term activities lasting minutes that don't involve stretching

C All site activities where a straight ladder cannot be used

D Getting on and off mobile towers

27.49

You have to carry out a job, over a few days, on the flat roof of a two-storey building, about 1 m from the edge of the roof, which has a very low parapet. You should:

A carry on with the job, provided you don't get dizzy with heights

B use a full body harness, lanyard and anchor while doing the job

C ask for double guard-rails and toe-boards to be installed to prevent you falling

D get your mate to do the work, while you hold on to them

27.50

You have been asked to install a number of ceiling-mounted air-conditioning units in a large open-plan area, which has a good floor. It is best to do this:

 using stepladders

B using scaffold boards and floor stands

C using packing cases to stand on

D from a mobile tower

F
27

28.1

When arriving at an occupied building, who or what should you consult before starting work to find out about any asbestos in the premises?

 A The person responsible for the building, to view the asbestos register

B The building receptionist

C The building logbook

D The building caretaker

28.2

Where might you find information on the safe way to maintain the services in a building?

A The noticeboard

B The safety officer

C The local Health and Safety Executive (HSE) office

D The health and safety file for the building

28.3

In the normal office environment what should be the hot-water temperature at the tap furthest from the boiler, after running it for one minute?

A At least 15°C

B At least 35°C

C At least 50°C

D At least 100°C

28.4

What should be the maximum temperature for a cold-water supply, after running it for two minutes?

A 10°C

B 20°C

C 35°C

D 50°C

28.5

What is required if there is a cooling tower on site?

A A formal logbook

B A written scheme of examination

C Regular visits by the local authority Environmental Health officer

D Inspections by the water supplier

28.6

Which TWO of the following are pressure systems?

A Medium and high temperature hot water systems at or above 95°C

B Cold water systems

C Steam systems

D Office tea urns

E Domestic heating systems

F
28

28.7

What is required before a pressure system can be operated?

(A) A written scheme of examination

(B) A permit to work

(C) City & Guilds certification

(D) A minimum of two competent persons to operate the system

28.8

On cooling tower systems, the water in the system should be:

(A) replaced annually

(B) chemically treated

(C) chilled

(D) drinking water

28.9

Which of the following should you NOT do when replacing the filters in an air-conditioning system?

(A) Put the old filters in a dustbin

(B) Follow a job-specific risk assessment and method statement

(C) Wear appropriate overalls

(D) Wear a respirator

28.10

After servicing a gas boiler what checks must you make by law?

(A) For water leaks

(B) For flueing, ventilation, gas rate and safe functioning

(C) The pressure relief valve

(D) The thermostat setting

28.11

Before adding inhibitor to a heating system what must you do?

(A) Check for leaks on the system

(B) Raise the system to working temperature

(C) Read the COSHH assessment for the product

(D) Bleed the heat emitters

28.12

When working in a riser how should access be controlled?

(A) By a site security operative

(B) By those who are working in it

(C) By the main contractor

(D) By a permit to work system

28.13

What is the correct action to take if natural gas is detected in an underground service duct?

(A) No action, if it is not harmful

(B) Evacuate the duct

(C) Carry on working but do not use electrical equipment

(D) Carry on working until the end of the shift

F
28

Answers: 28.7 = A   28.8 = B   28.9 = A   28.10 = B   28.11 = C   28.12 = D   28.13 = B

28.14

What should you think of first when planning to work in a confined space?

A) Whether the job has been priced properly

B) Whether sufficient manpower has been allocated

C) Whether the correct tools have been arranged

D) Whether the work could be done in some other way to avoid the need to enter the confined space

28.15

How should an adequate supply of breathable fresh air be provided in a confined space in which breathing equipment is not being worn?

A) An opening in the top of the confined space

B) Forced mechanical ventilation

C) Natural ventilation

D) An opening at the bottom of the confined space

28.16

What are the TWO MAIN safety considerations when using oxyacetylene equipment in a confined space?

A) The hoses may not be long enough

B) Unburnt oxygen causing an oxygen-enriched atmosphere

C) The burner will be hard to light

D) Wearing the correct goggles

E) The risk of a flammable gas leak

28.17

What precaution should be taken to protect against lighting failure in a confined space?

A) Remember where you got in

B) Ensure it is daylight when you do the work

C) Each operative should carry a torch

D) Secure a rope near the entrance and trail it behind you so that you can trace your way back

28.18

While working on your own and tracing pipework in a building, the pipes enter a service duct. You should:

A) go into the service duct and continue to trace the pipework

B) ask someone in the building to act as your second person

C) put on your personal protective equipment (PPE) and carry on with the job

D) stop work until a risk assessment has been carried out

F
28

28.19

Before working on electrically powered equipment, what is the procedure to make sure that the supply is dead before work starts?

- [A] Switch off and remove the fuses
- [B] Switch off and cut through the supply with insulated pliers
- [C] Test the circuit, switch off and isolate the supply at the mains board
- [D] Switch off, isolate the supply at the mains board, lock and tag out

28.20

When you are about to work on electrical equipment and the main isolator does not have a lock-out device, what is an alternative method of isolating the supply?

- [A] Detach the mains power supply
- [B] Place insulating material in the contactors
- [C] There is no alternative to a lock-out device for isolating the supply
- [D] Withdraw and retain the fuses and hang a warning sign

28.21

What would you use to find out whether a wall into which you are about to drill contains an electric supply?

- [A] A neon screwdriver
- [B] A cable tracer
- [C] A multimeter
- [D] A hammer and chisel

28.22

Which type of power drill is most suitable for fixing a run of pipework outside in wet weather?

- [A] Battery-powered drill
- [B] Drill with 110 volt power supply
- [C] Drill with 24 volt power supply
- [D] Any mains voltage drill with a power breaker

28.23

Temporary continuity bonding is carried out before removing and replacing sections of metallic pipework to:

- [A] provide a continuous earth for the pipework installation
- [B] prevent any chance of blowing a fuse
- [C] maintain the live supply to the electrical circuit
- [D] prevent any chance of corrosion to the pipework

28.24

What is the procedure for ensuring that the electrical supply is dead before replacing an electric immersion heater?

- [A] Switch off and disconnect the supply to the immersion heater
- [B] Switch off and cut through the electric cable with insulated pliers
- [C] Switch off and test the circuit
- [D] Lock off the supply, isolate at the mains board, test the circuit and hang a warning sign

F
28

28.25

What is used to reduce 240 volts to 110 volts on site?

A. RCD (residual current device)

B. Transformer

C. Circuit breaker

D. Step-down generator

28.26

What colour power outlet on a portable generator would supply 240 volts?

A. Black

B. Blue

C. Red

D. Yellow

28.27

What action should you take if a natural gas leak is reported in a closed area?

A. Ventilate the area and phone the gas emergency service

B. Establish whether or not it is safe to enter

C. Turn off the light

D. Wear safety footwear

28.28

Which of the following TWO actions should you take if a refrigerant leak is reported in a closed area?

A. Ventilate the area and extinguish all naked flames

B. Trace the leak and make a temporary repair

C. Establish whether or not it is safe to enter

D. Switch off the system

E. No action required as refrigerant gas is harmless

28.29

How is legionella transmitted?

A. Breathing contaminated airborne water droplets

B. Human contact

C. Dirty clothes

D. Rats' urine

28.30

What is the ideal temperature for legionella to breed?

A. Below 20°C

B. Between 20°C and 45°C

C. Between 45°C and 75°C

D. Between 75°C and 100°C

F
28

28.31

Which of the following is the most likely place to find legionella?

A Drinking water

B Hot water taps above 50°C

C Infrequently used shower heads

D The local river

28.32

When assembling a mobile access tower what major overhead hazard must you be aware of?

A Water pipes

B Cable trays

C False ceilings

D Suspended electric cables

28.33

You spill some oil on the floor and you do not have any absorbent material to clean the area. What should you do?

A Spread it about to lessen the depth

B Keep people out of the area and inform your supervisor

C Do nothing, it will eventually soak into the floor

D Warn other people as they tread through it

28.34

How should liquefied petroleum gas (LPG) cylinders be carried to and from premises in a van?

A In the back of the van

B In the passenger footwell of the van

C In a purpose-built container within the van

D In the van with all the windows open

28.35

When removing some panelling, you see a section of cabling with the wires showing. What should you do?

A Carry on with your work, trying your best to avoid the cables

B Touch the cables to see if they are live, and if so refuse to carry out the work

C Wrap the defective cable with approved electrical insulation tape

D Only work when the cable has been isolated or repaired by a competent person

28.36

When must a shaft or pit be securely covered or have double guard-rails and toe-boards installed?

A At a fall height of 1 m

B When there is a potential risk of anyone falling into it

C At a fall height of 2.5 m

D At a fall height of 3 m

F
28

Answers: 28.31 = C   28.32 = D   28.33 = B   28.34 = C   28.35 = D   28.36 = B

28.37

What should you do if, when carrying out a particular task, the correct tool is not available?

 A Wait until you have the appropriate tool for the task

B Borrow a tool from the building caretaker

C Use the best tool available in the toolkit

D Modify one of the tools you have

28.38

Who should be informed if a legionella outbreak is suspected?

A The Health and Safety Executive (HSE)

B The police

C A coroner

D The nearest hospital

28.39

When should the use of a permit to work be considered?

A All high risk work activities

B All equipment isolations

C At the beginning of each shift

D When there is enough time to complete the paperwork

28.40

Who should fit a padlock and tag to an electrical lock-out guard?

A Anyone working on the unit

B The engineer who fitted the lock-out device

C The senior engineer

D Anyone

28.41

You arrive on site and find the mains isolator for a component is switched off. What should you do?

A Switch it on and get on with your work

B Switch it on and check the safety circuits to see if there is a fault

C Contact the person in control of the premises

D Ask people around the building and if no-one responds, switch it on and get on with your work

28.42

When carrying out solvent welding on plastic pipework, what particular safety measure must you apply?

A The area must be well ventilated

B The supervisor must be present

C The area must be enclosed

D It must be done in daylight

F
28

28.43

Before starting work on a particular piece of equipment, who or what should you consult?

A. The machine brochure

B. The operation and maintenance manual for the equipment

C. The manufacturer's data plate

D. The store person

28.44

Which TWO of the following should a person who is going to work alone carry out to ensure their safety?

A. Register their presence with the site representative before starting work

B. Ensure their timesheet is accurate and countersigned

C. Make sure that somebody regularly checks that they are OK

D. Notify the site manager of the details of the work

E. Only work outside of normal working hours

28.45

To prevent unauthorised access to an unoccupied plant or switchgear room, what must you ensure?

A. That the access door is locked

B. That a sign is posted

C. That the power supply is isolated

D. That a person is posted to prevent access

28.46

What should folding stepladders be used for?

A. General access on site

B. Short-term activities lasting minutes that don't involve stretching

C. All work at height where a ladder cannot be used

D. Getting on and off mobile towers

28.47

When positioning and erecting a stepladder, which of the following is essential for its safe use?

A. It has a tool tray towards the top of the steps

B. The restraint mechanism is spread to its full extent

C. You will be able to reach the job by standing on the top step

D. A competent person has positioned and erected the steps

28.48

What is the first thing you should do after getting onto the platform of a correctly erected mobile tower?

A. Check that the brakes are locked on

B. Check the mobile tower to make sure that it has been correctly assembled

C. Close the access hatch to prevent falls of personnel, tools or equipment

D. Make sure that the tower does not rock or wobble

28.49

Before using a ladder at work, you notice that the maker's label says that it is a 'Class 3' ladder. What does this mean?

A It is for domestic use only and must not be used in connection with work

B It is of industrial quality and can be used safely

C It has been made to a European Standard

D It is made of insulating material and can be used near to overhead cables

F
28

Further information

Contents

Preparing for the case studies

Construction is an exciting industry. There is constant change as work progresses to completion.

As a result, the construction site is one of the most dangerous environments to work in.

Everyone on site working together can avoid many of the accidents that happen. The free film *Setting out* shows what you and the site must do to stay healthy and safe at work.

To watch the film:

 go online at *www.citb.co.uk/settingout.*

This film is essential viewing for everyone involved in construction, and should be viewed before sitting the CITB *Health, safety and environment test.*

The content of the film is summarised here. These principles form the basis for the behavioural case studies included in the test from Spring 2012.

It is advisable to watch the film. However, for those unable to view it or who want to refresh themselves on the content, the full transcript is provided for your information.

Part 1: What you should expect from the construction industry

Your site and your employer should be doing all they can to keep you and your colleagues safe.

Before any work begins, the site management team will have been planning and preparing the site for your arrival. It's their job to ensure that you can do your job safely and efficiently.

Five things the site you are working on must do:

- ☑ know when you are on site
- ☑ give you a site induction
- ☑ give you site-specific information
- ☑ encourage communication
- ☑ keep you up to date and informed.

Part 2: What the industry expects of you

Once the work begins, it's up to every individual to take responsibility for carrying out the plan safely.

This means you should follow the rules and guidelines as well as being alert to the continuing changes on site.

Five things you must do:

- ☑ respect and follow the site rules
- ☑ safely prepare each task
- ☑ carry out each task responsibly
- ☑ know when to stop (if you think anything is unsafe)
- ☑ keep learning.

Setting out – the transcript

What to expect from the industry and what the industry expects from you

Construction is an exciting industry. There is constant change as work progresses to completion.

As a result, the construction site is one of the most dangerous environments to work in.

Many accidents that occur on sites can be avoided. In this film you will find out what you and the site must do to stay healthy and safe at work.

Part 1: What to expect from the industry

When you arrive for your first day on a site, it will not be the first day for everyone.

Before any work begins, the site management team will have been planning and preparing the site for your arrival.

It is *their* job to ensure that you can do *your* job safely and efficiently.

So what are five key things the site should do for you?

1. Your site must know when you are on site

When you arrive you should be greeted and welcomed by someone on site. If you are not then make your presence known to site management.

You need to know who is in charge and they need to know who is working on, or visiting, their site.

You may be asked to sign in or report to someone in charge when you arrive. You should also sign out or let someone know if you are leaving.

2. Your site must give you an induction

Once you have introduced yourself, you will be given a site induction. This is a legal requirement to give you basic information so that you can work safely on site.

You may be asked to watch a video or look at a presentation. It is important that you understand what is said during the induction. If there is something you are not sure about, don't be afraid to ask for more information. If you have not been to the site for a while, you need to be sure that you are up to date. Check with site management whether you need a briefing or a further induction.

3. Your site must give you site-specific information

Whether you are just starting out or have decades of experience, every job is different. So it is very important that the site induction is specific to your site.

You should be told about any specific areas of danger and what site rules are in place to control these.

You will be told who the managers are on site and what arrangements are in place for emergencies.

You will also find out what to do it there is a fire or if you need to sound an alarm.

It may sound basic, but there must be good welfare facilities.

You must also be able to take a break somewhere that is warm and dry.

4. Your site must encourage communication

Evidence shows that there are fewer incidents and accidents on sites where workers are actively involved in health and safety. Your opinions and ideas are important so make them heard.

Whatever the size of your site there should be many opportunities to do this. For example, directly with managers through a daily briefing or through suggestion boxes. Managers should let you know how best to do this on your site.

5. Your site must keep you up to date

Construction sites are constantly changing and unplanned activities can be a major cause of accidents. The more up to date you are about what is happening on site, the more you can understand the dangers. The best sites keep their people informed on daily activities.

Is there a hazards board on your site – is it regularly updated?

Your site management should be telling you what's going on, on a regular basis.

Your site should be doing all it can to keep you and your colleagues safe. If it is not – say something and work together to make it better.

Part 2: What the industry expects from you

Site management will have planned ahead to make work on site as safe as possible for you. Once the work begins, it is up to every individual to take responsibility for carrying out the plan safely. This means following the guidelines set out and being alert to the continuing changes on site.

So what will the site expect from you?

1. You must respect site rules

Site rules are there to minimise the risk of particularly hazardous activities, such as moving vehicles and handling flammable substances.

Moving traffic on site is a major cause of accidents. Often rules will cover issues like walking through the site, where to park, and how to behave when you see a moving vehicle.

They may also tell you where you can smoke and remind you to tidy equipment away when it is not in use.

These might feel restrictive but they have been put in place for a reason. If that reason is not clear to you, ask for more information. Otherwise, follow the rules to stay safe.

2. You must safely prepare each task

Every task carried out on site is unique and will have its own dangers, for instance working at height or manual handling. The site management team will put in place a plan to avoid or minimise the dangers before the start of work. The plan may be written down in a risk assessment, a method statement or a task sheet.

This will tell you what to do, the skills needed, what to wear, what tools to use and what the dangers are.

You should contribute to the planning process using your experience and knowledge.

For example, before working at height, which is a high-risk activity, you must consider:

☑ Is the access suitable for what you need to do?

☑ Do you have the right tools?

☑ Do you have the right protective equipment, does it fit, and is it comfortable?

3. You must do each task responsibly

Once you are at work you must apply your training, skills and common sense to your tasks at all times.

For example, if you are building a wall, you may have to move heavy loads. But you must not put your health and body in danger. Make sure you move the load as safely as possible, as you should have been trained to do. And if you are not sure how to, then you must ask for advice.

Equally, it is important to be aware of the dangers to those around you. For example, if someone working with you is not wearing the correct PPE for a certain activity, tell them so.

Acting responsibly will benefit both you and your colleagues.

4. You must know when to stop

In our industry saying NO is not easy. We are fixers and doers. We CAN do.

However, a significant number of accidents on sites happen when people are doing things that they're not comfortable with.

For example, many workers have been harmed by not knowing how to identify asbestos. If you think there is any likelihood that there is asbestos present where you are working, you must stop work and seek advice.

If you are not properly trained, equipped, or briefed – or if the situation around you changes – the result could be an accident.

Trust your instincts. If things feel beyond your control or dangerous or, if you see someone else working unsafely, stop the work immediately and inform site management why you have done so.

You might prevent an injury or save a life. Your employer should be supportive if you do this because you have the right to say no and the responsibility to not walk by.

5. You must keep learning

If your job requires you to have specific training to enable you to do it safely then it is your employer's responsibility to provide it.

To really get the best out of your career, you should keep learning about developments in machinery, equipment, regulations and training.

This will not only give you greater confidence and understanding, it will ensure you remain healthy and safe.

In summary

Construction is so much more than bricks and mortar. The work we do improves the world around us. It's time for us to work together to build a safer and better industry.

Acknowledgements

CITB wishes to acknowledge the assistance offered by the following organisations in the preparation of the question banks:

- ☑ CITB (NI)
- ☑ Construction Employers Federation Limited (CEF NI)
- ☑ Construction Industry Confederation (CIC)
- ☑ Construction Plant-hire Association (CPA)
- ☑ Environment Agency
- ☑ Federation of Master Builders (FMB)
- ☑ Health and Safety Executive (HSE)
- ☑ Heating and Ventilating Contractors' Association (HVCA)
- ☑ Highways Agency
- ☑ Joint Industry Board for Plumbing Mechanical Engineering Services (JIB-PMES)
- ☑ Lift and Escalator Industry Association (LEIA)
- ☑ National Access and Scaffolding Confederation (NASC)
- ☑ National Demolition Training Group (NDTG)
- ☑ Scottish and Northern Ireland Joint Industry Board for the Plumbing Industry (SNIJIB)
- ☑ Strategic Forum for Construction Competence Working Group
- ☑ TunnelSkills
- ☑ UK Contractors Group (UKCG)
- ☑ Union of Construction, Allied Trades and Technicians (UCATT)
- ☑ Unison
- ☑ Unite